Lecture Notes in Mathematics

A collection of informal reports and seminars
Edited by A. Dold, Heidelberg and B. Eckmann, Zürich

Series: Mathematics Institute, University of Warwick
Adviser: D. B. A. Epstein

206

Proceedings of the Symposium on Differential Equations and Dynamical Systems

University of Warw
September 1968 –
Summer School, Jul

Edited by David Chillingworth,
University of Warwick, Coventry, Warwickshire/G.B.

Springer-Verlag
Berlin · Heidelberg · New York 1971

AMS Subject Classifications (1970): 34-02, 58-02, 57D30

ISBN 3-540-05495-2 Springer-Verlag Berlin · Heidelberg · New York
ISBN 0-387-05495-2 Springer-Verlag New York · Heidelberg · Berlin

Offsetdruck: Julius Beltz, Hemsbach

INTRODUCTION

These notes were taken from seminars and lectures given in the Mathematics Institute of the University of Warwick during the Symposium on Differential Equations and Dynamical Systems held between 1st September 1968 and 31st August 1969, and from lectures given in the Arts Lecture Theatre of the University during the Summer School in July 1969.

The organization of the Symposium and the Summer School would not have been possible without the generous support of the Science Research Council, to whom we are extremely grateful. The focus for the Symposium was the Mathematics Research Centre, which is partially supported by the Nuffield Foundation, to whom we are also very grateful. We extend our particular thanks to Professor L. Markus, the Nuffield Professor, for his energies in running the Symposium so successfully.

The notes are presented in approximately chronological order, with the exception of the Summer School lectures which are placed at the end and the talks from a sub-seminar on Foliations which are collected together. Some of the lectures were expositions of work already published, while others announced results which were obtained during the course of the Symposium. The level of familiarity of the audience with the subject matter which was assumed by the speakers varied considerably, and so terms left undefined in one lecture may frequently be found explained in another. To avoid repetition, some standard notation which is used throughout (except where stated otherwise) is collected on page XI. References [1] indicate the references at the end of each lecture, and references (2) refer to other lectures in these Proceedings. Many of the references given as "to appear" have now appeared; in particular, the Proceedings of the A.M.S. Summer Institute on Global Analysis, Berkeley, 1968 are now in print. A good general background to many of the topics and their motivations is the survey article Differentiable dynamical systems (Bull. A.M.S. 73 (1967) 747-817) of S. Smale. The books Arnol'd and Avez - Problèmes Ergodiques de la Mécanique Classique (Gauthier-Villars, Paris, 1967. (English version: Benjamin, 1968)) and Abraham and Marsden - Foundations of Mechanics (Benjamin, 1967) are also useful references, especially for material on ergodic theory and Hamiltonian systems respectively.

The four lecture-courses given during the morning sessions of the Summer School by L. Markus, S. Smale, R. Thom and E.C. Zeeman are not included here.

I would like to express my thanks to all the contributors for their assistance in the writing and checking of these lecture notes, and especially to the speakers at the Summer School who helped greatly by providing their own abstracts of their lectures. I am also grateful to Lynn Herbert and Anne Finch who battled on in the hot summer of 1969 to type the first draft of these Proceedings, and to Mrs. M. Matheson who cheerfully typed the whole of the final version.

I regret that it has taken so long for these notes to be published: the causes involve difficulties in communication during absences from Warwick, illnesses, accidents and the British postal workers' strike - all, however, allowed to produce more difficulty than necessary by the insuffiency of *yang* in my own organization. Inevitably, some of the results announced are out of date. Nevertheless, I hope that the collection in one volume of these notes on 86 lectures in many different aspects of dynamical systems may be of use both to those who are approaching the subject for the first time and are looking for ideas and directions to follow, as well as to those who already work in this field and are interested to note the ways in which their own branch relates to the subject as a whole.

March 1971 David Chillingworth.

CONTENTS

SYMPOSIUM LECTURES

SEMINAR ON FOLIATIONS

SUMMER SCHOOL LECTURES

SOME STANDARD NOTATION

Z	integers
R	real numbers
Q	rational numbers
$\mathbb{C}$	complex numbers
R^n	real euclidean n-space
$\mathbb{C}^n$	complex n-space
S^n	n-sphere
T^n	n-torus = $S^1 x \ldots x S^1$ (n times)
$RP^n(CP^n)$	real (complex) projective n-space
SL(n,R) (SO(n))	group of n×n real (orthogonal) matrices with determinant +1
M	a C^∞ finite-dimensional manifold
TM(T*M)	tangent bundle (cotangent bundle) of M
$\mathrm{Diff}^r(M)$ (Diff(M))	group of C^r diffeomorphisms (C^∞ diffeomorphisms) of M
Ω or $\Omega(f)$	non-wandering set of $f \in \mathrm{Diff}^r(M)$ or Diff(M)
C(X)	continuous complex-valued functions on X
$\mathcal{L}_p(Y)$	p^{th}-power-integrable functions on Y
∂	boundary

(1) <u>On the measure-preserving flow on the torus</u> <u>T. Saito</u>

Consider the differential equation on the torus T^2 induced by the system $\frac{dx}{dt} = X(x,y)$, $\frac{dy}{dt} = Y(x,y)$ on the plane R^2, where X and Y are periodic in x,y with period 1. We assume: (1) X,Y are both C^1 (and so solutions exist and are unique), (2) there are no critical points (i.e. where X = Y = 0), (3) the resulting flow is measure-preserving (which is the case if and only if $\frac{\partial X}{\partial x} + \frac{\partial Y}{\partial y} = 0$) and ask: <u>Find necessary and sufficient conditions for every orbit to be dense on T^2</u>.

Let w = Ydx - Xdy; then w is exact so there exists H(x,y) such that $\frac{\partial H}{\partial x} = Y$, $\frac{\partial H}{\partial y} = -X$. Although H(x,y) is constant on each orbit and single-valued in R^2, it may not even be single-valued on T^2 (i.e. H may not be periodic). We use H to prove the following:

<u>THEOREM 1</u>. Let $X = a+\sum a_{mn} e^{2\pi i(mx+ny)}$, $Y = b+\sum b_{mn} e^{2\pi i(mx+ny)}$ $((m,n) \neq (0,0))$ be the Fourier expansions of X and Y. Then every orbit is dense on $T^2 \longleftrightarrow ab \neq 0$ and a/b is irrational [1].

<u>Proof</u>. It is easy to verify that H(x+1,y) - H(x,y) is a constant α, say, and similarly $H(x,y+1) - H(x,y) = \beta$. Then, from the fact that $H(x,y) - \alpha x - \beta y = F(x,y)$ is periodic and $\frac{\partial H}{\partial x} = Y$, $\frac{\partial H}{\partial y} = -X$ it follows that $\alpha = b$ and $\beta = -a$; thus $H = bx-ay+F$ where F is single-valued on T^2. If a = b = 0 then F is constant on each orbit, so the existence of a dense orbit would imply F constant everywhere, which is impossible since then $X \equiv Y \equiv 0$. Assume $b \neq 0$. (1) <u>a/b rational</u>. If a = r/k, b = s/k ($r,s \in Z$) then $G = \cos 2\pi(rx-sy+kF)$ is single-valued on T and constant on each orbit, so as above no dense orbit exists. (2) <u>a/b irrational</u>. There is no periodic orbit since $bp-aq = 0$ ($p, q \in Z$) $\rightarrow p = q = 0$ and so any closed

orbit would lift to R^2 and hence have a critical point inside it. (If X,Y are C^2 our result now follows by [2]). Let $\Gamma(\cong S^1)$ be a section of the flow and let $f:\Gamma\to\Gamma$ be the Poincaré map induced by the flow [2]. Since the derivative of H along Γ never vanishes and $\int_\gamma dH = \int_{f(\gamma)} dH$ for any arc $\gamma \subset \Gamma$, f is topologically equivalent to a measure-preserving homeomorphism of S^1, which has no periodic points since the flow has no periodic orbits. Hence f is equivalent to $\theta \mapsto \theta + 2\pi\alpha$ where α is irrational, and so every orbit on T^2 is dense.

THEOREM 2. The flow is ergodic $\Longleftrightarrow$ $ab \neq 0$ and a/b is irrational.

The proof of Theorem 2 may be found in [1].

REFERENCES:

[1] Saito, T. On the measure-preserving flow on the torus, J. Math. Soc. Japan 3 No. 2 (1951) 279-284.

[2] Siegel, C.L. Note on differential equations on the torus, Ann. Math. 46 (1945) 423-428.

(2) Breaking of waves

E.C. Zeeman

In his book "Stabilité Structurelle et Morphogenèse" (Benjamin, 1969) Thom proves:

THEOREM. Up to diffeomorphism there are seven elementary catastrophies.

In other words, we have seven standard geometric models of discontinuity, and any elementary discontinuity arising in nature is diffeomorphic to one of these. Elementary means having a finite number of types of state in the neighbourhood. The breaking wave is elementary, with five states: water[1], air[2], surface[3], crest[4] and first appearance of crest.[5]

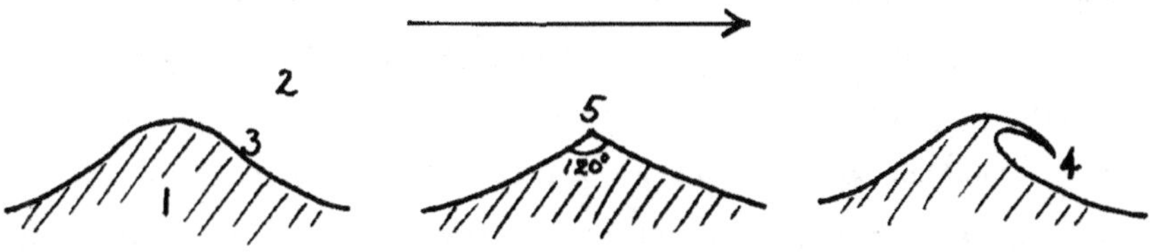

If we assume there is an implicit mathematical model for the wave, with implicit physical variables minimising some potential, then Thom's theorem applies. The relevant catastrophe is number five (hyperbolic umbilic), the other six being of wrong dimension or shape. The standard model of the hyperbolic umbilic comprises a surface $S \subset R^3$ given by $b^2 = (a^2 - 4\ ac^2 + c^4) - \frac{2c}{\sqrt{3}}(c^2 - 2a)^{3/2}$. We explain below how this formula is obtained. The cross-sections (c = constant) of S are rounded ($c < o$), right-angled ($c = o$) and cusp ($c > o$). For $c > o$ the crest line is $a = -c^2$, $b = o$. Just below the crest, if $a = -c^2(1 + \varepsilon)$ then $b = \pm \frac{c^2}{3} \varepsilon^{3/2} + O(\varepsilon^2)$ which indicates the cusp.

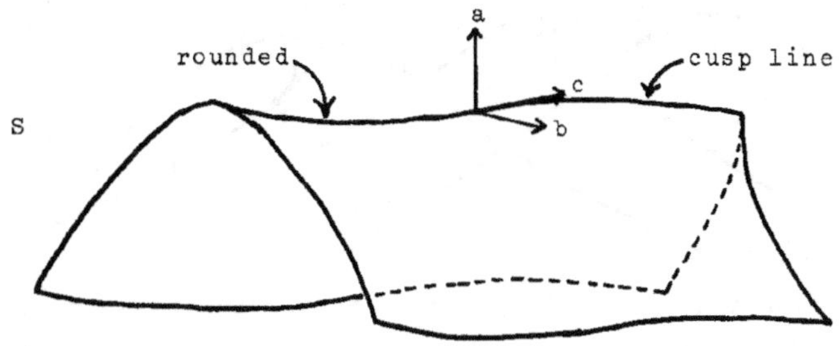

There are two ways of applying the result. (1) Given a wave breaking diagonally onto the shore, there is a diffeomorphism $R^3 \to R^3$ (space) throwing S locally (i.e. near the origin) onto the water surface. (2) Given a wave parallel to the shore, it suffices to look at cross-sections: we have a diffeomorphism $\phi : R^3 \to R^2 \times R^1$,

$\mathbb{R}^2$(space) × R^1 (time), throwing S locally onto the water surface. ϕ preserves the cusps, and near O we can take the linear approximation $d\phi$. If $d\phi^{-1}$ (time axis) is in the (a,c) - plane we have a standing wave breaking upwards, or (by symmetry) a wave breaking against a jetty. If $d\phi^{-1}$ (time axis) has a b-component we have a travelling wave.

Consider the second case. Whatever the angle of break, there is a cusp immediately after. In fact, Stokes showed the angle to be 120°, and shallow water theory shows the wave symmetrical before break - both results confirmed by observation. Now, theory also shows that the particle at the top of the wave travels faster than the wave (in a sine-wave twice as fast). Combining this with Thom's cusp, we deduce that just after the break these particles fall freely in a parabola under gravity. There will be a point p on the leading face above which parabolic fall occurs, and below which the classical wave continues.

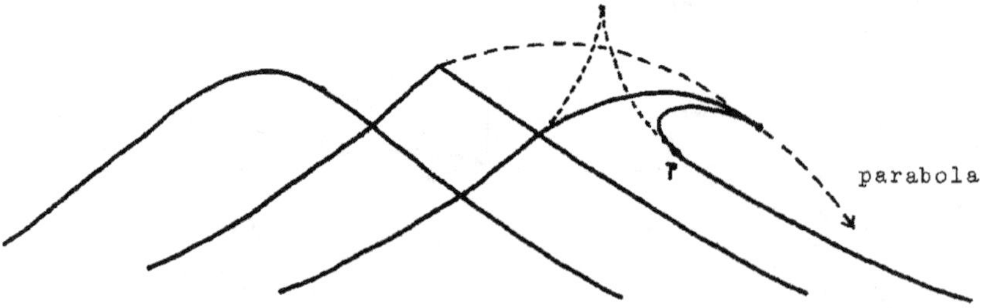

Thus, one can predict qualitatively the parabolic shape with cusp edge. (Near the edge itself the model no longer continues to be valid: a further non-elementary catastrophe takes place causing a frill, followed by parabolic umbilics breaking off droplets.)

The above description agrees with the observation (first pointed out to me by H. Peregrine) that a thin jet-stream seems to shoot off the top of the wave. Although it happens rather quickly for the eye to be sure, the phenomenon shows clearly in photographs that any amateur can take. There are two optical illusions to beware of: (1) the rising curvature of the leading face looks steeper than it is from the front, (2) the jet-stream looks thicker than it is from above. Both illusions can be avoided by placing the eye near water-level so as to look into the "hole" of a diagonally breaking wave. The incompressibility of water causes $d\phi$ (c-axis) to have a vertical component, so the parabola has an initial upward component. This component, and hence the shape of the parabola, depend on the size and speed of the wave and slope of the shore.

Summarising, to explain the shape of the forward-breaking water wave, one has to combine classical hydrodynamics with Thom's catastrophe theory; what is now needed is an explicit identification of the catastrophe potential V in terms of classical physical variables. Thom's theory is also applicable to backward-breaking waves, as for example meteorological waves in the front between polar and temperate air-masses, the cusp being the well known occlusion between the warm and cold fronts of a depression. We conclude by showing how to obtain the all important surface S.

Construction of S. The hyperbolic umbilic is controlled by a generic cubic potential in two variables, with hyperbolic discriminant, for which we can choose (after Thom) $V = x^3 + y^3$ with unfolding

$$U = x^3 + y^3 + (a+b)x + (a-b)y + 2cxy.$$

The unfolding is merely a convenient $R^3 = \{(a,b,c)\}$ of potentials, transversal to the bifurcation stratum containing V. It is the intersection of R^3 with the bifurcation set that gives S, as follows.

For certain values of (a,b,c) the potential U has one minimum (representing the state of water) and for other values U has no minimum (representing air). The surface S separating these two states is given by eliminating x,y from the three equations:

$$U_x = U_y = \begin{vmatrix} U_{xx} & U_{xy} \\ U_{yx} & U_{yy} \end{vmatrix} = 0.$$

The vanishing of U_x, U_y gives the minimum, and the vanishing of the discriminant gives where the minimum coalesces with a saddle. We obtain

$$4\left((a+b)^2 + (a-b)c^2\right)\left((a-b)^2 + (a+b)c^2\right) = (a^2-b^2-c^4)^2,$$

and solving for b we obtain

$$b = \pm\sqrt{(a^2-4ac^2+c^4) \pm \frac{2c}{\sqrt{3}}(c^2-2a)^{3/2}}.$$

Choosing the minus sign of the second ± eliminates the dual surface related to maxima, which is irrelevant in this context.

(3) Group extensions of discrete dynamical systems W. Parry

A discrete dynamical system (X,S) consists of a homeomorphism S on a compact metric space X. If G is a compact group acting freely on X and commuting with S, then S induces a homeomorphism T on Y = X/G : we say (X,S) is a G-extension of any system equivalent to (Y,T). There is a general problem of determining which properties of (Y,T) lift to (X,S): here we consider the properties of minimality and unique ergodicity. We assume G is abelian.

Definitions:

(i) (Y,T) minimal means the only closed sets of Y invariant under

T are Y and $\emptyset$.

(ii) (Y,T) uniquely ergodic (u.e.) means there exists only one normalized ($\mu(Y)=1$) Borel measure μ invariant under T. This is equivalent to the existence of only one ergodic measure μ (i.e. such that the only Borel sets E with $TE = E$ satisfy $\mu(E) = 0$ or 1), and to:

$$\lim_{n\to\infty} \frac{1}{n} \sum_{i=o}^{n-1} f(T^i y) = \int f d\mu$$

for all $y \in Y$ and all $f \in C(Y)$.

Clearly, if (Y,T) is u.e. (with ergodic measure μ) and μ(open sets) > 0, then (Y,T) is minimal. If $\tilde{m}$ is a measure for Y then the natural measure m for X is that defined by $\int f dm = \int (f(gx)dg)d\tilde{m}$.

THEOREM 1 [1]. If (Y,T) is u.e. (measure $\tilde{m}$) and (X,S) is ergodic with respect to the natural measure m, then (X,S) is u.e..

Proof. Let μ be any ergodic measure for (X,S): we show $\mu = m$. For $\nu = \mu$ or m let E_ν =

$$\{x \mid \lim_{n\to\infty} \frac{1}{n} \sum_{i=o}^{n-1} f(S^i x) = \int f d\nu \text{ for all } f \in C(X)\}.$$

Then $\nu(E_\nu) = 1$ by the Ergodic Theorem. Define $\tilde{\mu}$ by $\tilde{\mu}(F) = \mu(\pi^{-1}F)$: since Y is u.e. we have $\tilde{\mu} = \tilde{m}$. Now $\tilde{\mu}(\pi E_\mu) = \mu(\pi^{-1}\pi E_\mu) \geq \mu(E_\mu) = 1$ and similarly $\tilde{m}(\pi E_m) \geq 1$. Hence there is $x \in E_\mu$ such that $gx \in E_m$, and then $x \in E_m$ so

$$\int f d\mu = \lim_{n\to\infty} \frac{1}{n} \sum_{i=o}^{n-1} f(S^i x) = \int f dm$$

for all $f \in C(X)$. Therefore $\mu = m$.

It is possible to give equations involving S and T which, when never satisfied by any $f \in C(Y)$, imply the lifting of minimality or u.e. from (Y,T) to (X,S). However, in any given case it is difficult to verify whether the equations are satisfied or not.

If $g \in G$ then the translate gS defines another system (X, gS).

THEOREM 2 [3]. Let X be connected. The set $\{g \mid (X, gS)$ has the same properties (i), (ii) as $(Y, T)\}$ is a dense G_δ-set in G.

Define S to be stable if gS is conjugate to S for all g in a neighbourhood of $e \in G$. Then clearly S stable $\rightarrow$ properties lift to (X, S). By lifting to an n-torus the system (S^1, T) on the circle, where T is given by $T(x) = x + \alpha \bmod 1$ (α irrational), and using Theorem 1, it is easy to prove Weyl's theorem: If $f \in C([0,1])$ then

$$\lim_{n\to\infty} \frac{1}{n} \sum_{k=0}^{n-1} f(p(k)) = \int_0^1 f dx,$$

where $p(k) = a_q k^q + \ldots + a_1 k + a_0$ and some a_i $(i > 0)$ is irrational, [2].

REFERENCES:

[1] Furstenberg, H. Strict ergodicity and transformations of the torus, Amer. J. Math. 83 (1961) 573-601.

[2] Parry, W. Compact abelian group extensions of discrete dynamical systems, (to appear).

[3] Hahn, F. On affine transformations of compact abelian groups, Amer. J. Math. 85 (1963) 428-446.

(4) Conditions for integrability of certain equations J.Kurzweil

We consider systems of total differential equations of the form

$$(1)\ \frac{\partial y_i}{\partial z_j} = H_{ij}(y,z),\ i = 1,2,\ldots,m,\quad j = 1,2,\ldots,n,$$
$$y = (y_1,\ldots,y_m),\ z = (z_1,\ldots,z_n),$$

or, briefly, $dy = Hdz$, and give a new proof of the following result due to P. Hartman [1,2,3]:

Assume that H_{ij} are continuous. For a differential 1 - form ξ and 2 - form ω (both with continuous coefficients) write $\omega = d\xi$ if the Stokes formula $\int_{\partial B} \xi = \int_B \omega$ holds for all rectangles B. For any such ξ there exists at most one $d\xi$. For a continuous matrix $A(y,z) = (A_{ij}(y,z))$ put $\xi_i = \sum_j A_{ij}(dy_j - \sum_k H_{jk}dz_k)$, i.e. $\xi = A(dy-Hdz)$.

THEOREM. (1) is integrable if and only if there exists an A such that ξ_i have continuous derivatives for $i = 1,2,\ldots,m$ and $\xi_1 \wedge \xi_2 \wedge \cdots \wedge \xi_m \wedge d\xi_i = 0$ for $i = 1,2,\ldots,m$.

REFERENCES:

[1] Hartman, P. Ordinary Differential Equations, J. Wiley and Sons, New York, 1964.

[2] Hartman, P. On exterior derivatives and solutions of ordinary differential equations, Trans. A.M.S. 91 (1959) 277-292.

[3] Hartman, P. On uniqueness and differentiability of solutions of ordinary differential equations, Proc. Symposium on Nonlinear Problems, Madison (Wis) 1963, 255-259.

(5) Formalisme Lagrangien C. Godbillon

Les équations de Lagrange se transformant de façon contravariante. Elles sont donc associées à un champ de vecteurs sur l'espace tangent T(M) à la variété de configuration M. La construction de ce champ de vecteurs repose sur une étude du calcul différentiel de cet espace tangent.

Si $(q_1,\ldots,q_m)$ est un système de coordonnées locales sur un ouvert U de M,

$$(q_1,\ldots,q_m,\ \dot{q}_1=dq_1,\ldots,\dot{q}_m=dq_m)$$

est un système de coordonnées locales sur l'ouvert $p^{-1}(U)$ de T(M).

1. Calcul différentiel sur T(M).

Proposition: Il existe une suite exacte

$$0 \rightarrow p^*T(M) \overset{H}{\rightarrow} T(T(M)) \overset{K}{\rightarrow} p^*T(M) \rightarrow 0.$$

L'application v = H.K est l'endomorphisme vertical du deuxième fibré tangent T(T(M)). On a $v^2 = 0$.

Proposition: Il existe une dérivation i_v de degré 0 de l'algèbre Λ (T(M)) des formes différentielles sur T(M) et une seule ayant les propriétés suivantes:

$$i_v f = 0,\ i_v df = v^* df,\ f \in \mathfrak{D}(T(M)).$$

Proposition: Le crochet $d_v = [i_v,d] = i_v d - d i_v$ est une antidérivation de degré + 1 de Λ (T(M)).

d_v est la différentiation verticale dans Λ (T(M)). On a $d_v d_v = 0$ et $d_v d = - d d_v$.

Localement $\quad d_v f = \Sigma_i \dfrac{\partial f}{\partial \dot{q}_i} dq_i$

$$d_v dq_i = d_v d\dot{q}_i = 0.$$

Définition: L'algèbre des formes différentielles semi-basiques est l'image de l'endomorphisme v^* de Λ (T(M)).

Une forme de Pfaff semi-basique s'écrit localement

$$\Sigma_i a_i(q_1,\ldots,q_m,\ \dot{q}_1,\ldots,\dot{q}_m)\ dq_i.$$

2. Systèmes mécaniques.

Définition: Un système mécanique est un triplet $\mathfrak{M} = (M,T,\pi)$ où

M^m <u>est une variété différentiable</u>,

$T : T(M) \to R$ <u>une fonction différentiable</u>,

π <u>une forme de Pfaff semi-basique sur</u> $T(M)$.

La forme différentielle $\omega = dd_V T$ est la <u>forme fondamentale</u> de $\mathfrak{M}$.

Le système mécanique $\mathfrak{M}$ est <u>régulier</u> si ω est une forme symplectique ($\omega^m \neq 0$).

<u>Proposition</u>: <u>Soit</u> $\mathfrak{M} = (M,T,\pi)$ <u>un système mécanique régulier</u>. <u>Il existe un champ de vecteurs</u> X <u>sur</u> $T(M)$ <u>et un seul tel que</u>

$$i_X\omega = d(T - V.T) + \pi.$$

Ici $i_X\omega$ est le produit intérieur de ω par X et V le champ de vecteurs de Liouville sur $T(M)$ (V engendre le groupe à un paramètre des homothéties de $T(M)$).

<u>THEOREME</u>: X <u>est une équation différentielle du second ordre sur</u> M.

Localement X est donné par les équations de Lagrange.

<u>Définition</u>: <u>Le système mécanique</u> $\mathfrak{M}$ <u>est conservatif si</u> π <u>est une forme fermée</u>.

Le champ X est alors un champ de vecteurs hamiltonien associé à la forme symplectique ω.

<u>Définition</u>: <u>Le système mécanique</u> $\mathfrak{M}$ <u>est lagrangien s'il existe une fonction</u> U <u>sur</u> M <u>telle que</u> $\pi = p^* dU$.

$L = T + U \circ p$ est le <u>lagrangien</u> de $\mathfrak{M}$.

$H = V.T - T - U \circ p$ le <u>hamiltonien</u> de $\mathfrak{M}$.

Dans ce cas X est le système hamiltonien défini par

$$i_X\omega = - dH.$$

(6) Continuous flows on the plane A. Beck

We consider a flow $\phi : R \times X \to X$ on a metric space X and ask the question: What can the fixed-point set $F(\phi)$ of ϕ look like? When X is the plane it is easy to construct a flow having any given closed set F as fixed-point set (e.g. take ϕ to be the solutions of $\dot{x} = d(x,F)\ \underline{v}_o$ where d is the distance function and $\underline{v}_o$ is any fixed vector), and in fact there is the following general result:

THEOREM 1. If ϕ_o is a continuous flow on the metric space X, and F is any closed set containing $F(\phi_o)$, then there exists ϕ_1 with $F(\phi_1) = F$, [1].

Returning to the plane E^2, we observe that the flow constructed above may in general have an infinite number of stagnation points (i.e. points $y \in F(\phi)$ such that y = end point of some orbit θ but $y \notin \theta$) and thus not readily represent a physical system; hence we ask: What can $F(\phi)$ look like if ϕ has only a finite number n of stagnation points? Some answers are contained in the following wider theorem [2,3]:

THEOREM 2. I. Suppose (i) all orbits of ϕ are either points or circles OR (i)' ϕ has no stagnation points. Then

$$F(\phi) = E^2 - \bigcup_\alpha A_\alpha$$

where the A_α are homeomorphic to open annuli.

II. Suppose (ii) all orbits of ϕ are closed sets OR (ii)' ϕ has just one stagnation point, which is set at infinity. Then

$$E^2 - [F(\phi) \cup G(\phi)] = \bigcup_\alpha A_\alpha$$

where $G(\phi)$ is the set of points whose orbits are non-compact.

Notes: (1) The actual flows in (i), (ii)' are very different from those of (i), (ii). (2) The situation becomes enormously complicated when $n = 2$, but then not much more so for $n > 2$, [3]. (3) The frontier of A_α may look nothing like two circles.

THEOREM 3. Given a closed set $F \subset E^2$ it is possible to find ϕ with countably many stagnation points so that $F(\phi) = F$, [3].

REFERENCES:

[1] Beck, A. On invariant sets, Ann. Math. 67 (1958) 99-103.

[2] Beck, A. Plane flows with closed orbits, Trans. A.M.S. 114 (1965) 539-551.

[3] Beck, A. Plane flows with few stagnation points, Bull. A.M.S. 71 (1965) 886-890.

(7) Non-linear cubic differential equations — T.V. Davies

Sufficient conditions for the existence of a centre (i.e. for a continuum of closed integral curves) at $x = y = 0$ for the differential equation

$$\frac{dy}{dx} = \frac{-x + a_2x^3 + 3b_2x^2y + 3c_2xy^2 + d_2y^3}{y + a_1x^3 + 3b_1x^2y + 3c_1xy^2 + d_1y^3} = \frac{-x + Q_3}{y + P_3}$$

can be shown to be

Case I
$$a_1+b_2+c_1+d_2 = 0$$
$$(a_1+3b_2+2d_2)(b_1+c_2) + (a_2+d_1)(a_1+b_2) = 0$$
$$(a_1+b_2)^2\{(a_1+b_2)(a_2+d_1) - (a_1-3b_2)(b_1+c_2)\} =$$
$$2(b_1+c_2)^2\{a_2(a_1+b_2)+a_1b_1-2a_1c_1+3b_1b_2\}$$

Case II
$$a_1 = -6b_1-5d_2, \qquad b_1 = 5a_2+4c_2$$
$$c_1 = 5b_2+4d_2, \qquad d_1 = -4a_2-3c_2$$
$$(9b_2+7d_2)(3b_2+d_2) + 4a_2(4a_2+3c_2) = 0.$$

When the centre is perturbed it appears that the perturbed

equation

$$\frac{dy}{dx} = \frac{-x + \lambda y + Q_3}{y + \lambda x + P_3}$$

possesses at most five limit cycles in a neighbourhood of the above 'centre' manifolds.

(8) Mathematical theory of general systems M.D. Mesarović

The mathematical theory of general systems is concerned with unification of, and foundations for, various branches of applied mathematics dealing with mathematical models of information processing and decision-making systems, such as dynamical systems theory, control theory, automata theory, etc.

The theory is being developed in an axiomatic way in the following manner: Intuitive concepts are formalized in as weak a mathematical structure as the proper interpretation would allow and the consequences of additional assumptions required for specialized applications are then investigated.

Starting from the notion of a general system, S, as a relation on abstract sets, two tracks are pursued:

1) A binary function, C, defined on a subset of $S \times S$ is introduced: the partial algebra $\{S,C\}$ is then used as the basis for an algebraic theory of systems;

2) S is defined as $S \subset A^T \times B^T$ where T is a linearly ordered set while A and B are arbitrary; S is then used as a basis for the development of the theory of general time systems.

Several notions, such as state transition, state space, controllability, stability, are formalized and some results concerning the relationship between these concepts and its further application in specialized situations are presented.

In particular, a decomposition theorem which gives the conditions for the state-determined, causal, representation of a system is presented, with its application in dynamical systems theory and automata theory.

(9) Mathematical theory of multi-level systems M.D. Mesarović

A control (or decision-making) system which consists of a family of hierarchically arranged control subsystems is termed a multi-level system. It has been shown that many complex industrial, biological and organizational systems can be modelled mathematically in such a way.

A central problem in the theory of multi-level systems is the interrelationship between levels, and in particular exchange of information and decisions. To develop a mathematical theory of interlevel relationship the simplest type of multi-level system, namely a two-level system with a single unit on the first level, has been considered; the decision problem of the second level unit in such a system is termed co-ordination.

A mathematical theory of co-ordination for systems described in Banach spaces is presented. Theorems giving necessary and sufficient conditions for optimal co-ordination of abstract dynamical systems are presented and their application to systems described by differential equations is discussed.

(10) Geometric elements in the theory of transformations of ordinary second-order linear differential equations
O. Borůvka

The basic problem in the theory of transformations of second-order linear differential equations formulated by Kummer in 1834 is the following:

Given two such equations (1) $y'' = q(t)y$, (2) $Y'' = Q(T)Y$

defined on some intervals of the real line, how can one find functions $w(t)$, $X(t)$ (with $w(t) \neq 0$, $X'(t) \neq 0$) such that $y(t) = w(t).Y[X(t)]$ is a solution of (1) whenever $Y(T)$ is a solution of (2)?

In fact, it turns out that (under some supplementary conditions) w, X satisfy the above if, and only if, $w \equiv C/\sqrt{X'}$ (C = constant) and X is a solution of the equation

$$- \{X,t\} + Q(X)X'^2 = q(t)$$

where $\{X,t\}$ denotes the Schwarzian derivative of X at t.

Some important notions which occur in the theory of transformations are the following. Let u,v be two linearly independent solutions of (1); then the <u>first and second phases</u> with respect to the basis (u,v) are any real functions $\alpha(t)$, $\beta(t)$ continuous where (1) is defined and satisfying (for $v,v' \neq 0$) $\tan \alpha = u/v$, $\tan \beta = u'/v'$ respectively. The function $\theta = \beta - \alpha$ is called a <u>polar function</u> of the basis (u,v).

Let $q(t) < 0$ for all t. Take an arbitrary t and suppose $u(t) = 0$. Define $\phi_n(t) = n^{th}$ zero of u following t, $\chi_n(t) = n^{th}$ zero of u' following t. Now instead suppose $v'(t) = 0$, and let $\psi_n(t)$, $\omega_n(t)$ be the nth zeros of v', v (respectively) following t. Define also $\phi_{-n}(t)$ etc. by replacing 'following' by 'preceding'. Then the ϕ_ν ($\nu = \pm 1, \pm 2, \ldots$) etc. are called the <u>central dispersions</u> with index ν, and do not depend on the choices of u and v. When $\nu = +1$ they are called <u>basic central dispersions</u>. There exist numerous formulae which relate all these quantities, and give an effective analytic apparatus for solving various problems in the theory of differential equations.

An equation (3) $Y''+AY'+BY = 0$ defines (up to non-singular homogeneous linear transformation) a planar curve, by the locus of $(U(t), V(t))$ where U,V are two independent integrals. Conversely, if U,V are C^2 functions with $UV'-VU' \neq 0$ then there is an equation (3) with U,V as independent integrals. The curve $\mathcal{C}$ given by (U,V) can be reparametized so that (3) takes the form (1), and assuming $q < 0$ the central dispersions can be interpreted geometrically: the points on $\mathcal{C}$ given by t and $\phi_\nu(t)$ lie on the same straight line through the origin, the tangents at points given by t and $\psi_\nu(t)$ are parallel, and so on.

The above can be used to attack problems of the following type:

(a) Which differential equations satisfy $\phi_1(t) = \psi_1(t)$?

(b) Which satisfy $\phi_1(t) = \Phi(t)$, for given Φ?

Problem (a) has been thoroughly studied, and it is known that $\phi_1(t)$ must be of the form $ct+k$ $(c>0)$, and $\mathcal{C}$ has polar equation of the form $r = C^\alpha F(\alpha)$ where $C > 0$ and $F(> 0)$ is a C^2 periodic function satisfying a certain differential inequality. For (b) it is known, for instance, that if Φ is C^3 with $\Phi(t) > t$, $\Phi'>0$ and $\Phi \to \pm\infty$ as $t \to \pm\infty$ (resp.) then there exists an uncountable set of equations (1) with $\phi_1 = \Phi$. The corresponding functions q can be given quite explicitly, especially in the case $\Phi(t) = t+k$.

More detailed information is available in [1]; further geometric applications of the theory of transformations have been made in [2].

REFERENCES:

[1] Borůvka, O. Lineare Differentialtransformationen 2. Ordnung, Deutscher Verlag der Wissenschaften, Berlin, 1967.

[2] Guggenheimer, H. (To appear in Archivum Mathematicum, (Brno)).

(11) Dynamical systems on an n-torus T. Saito

A dynamical system on the n-torus T^n is defined by equations

$$\frac{dx_j}{dt} = X_j(x), \qquad j = 1, \ldots, n$$

where each X_j is a C^1 function with period 1. Assuming the resulting flow to be measure-preserving (i.e. $\sum_{j=1}^{n} \frac{\partial X_j}{\partial x_j} = 0$) we ask: under what conditions is the flow minimal (i.e. every orbit dense)? When n = 2 the problem has already been solved [1]; in view of that solution we might expect in general that the flow is minimal provided $a_1, \ldots, a_n$ are linearly independent over the integers, where

$$X_j = a_j + \Sigma a^{(j)}_{m_1 \ldots m_n} \exp 2\pi i \ (m_1 x_1 + \ldots + m_n x_n).$$

THEOREM. If (1) $a_1, \ldots, a_n$ are linearly independent over the integers and (2) every orbit of the system is Liapunov stable in both directions, then there exists a homeomorphism $f : T^n \to T^n$ taking orbits into straight lines with slopes $(a_1, \ldots, a_n)$, and therefore the flow is minimal.

Proof. Lift the system to R^n and let $x_j(t;\underline{x}_0)$ denote the j-component of the solution stating at $\underline{x}_0 = (x_{10}, \ldots, x_{n0})$. Uniqueness of solutions shows

$$x_j(t;\underline{x}_0) = x_{j0} + f_j(t;\underline{x}_0)$$

where f_j is periodic. Then

$$f_j(t;\underline{x}_0) = \alpha_j(t) + \Sigma \alpha^{(j)}_{m_1 \ldots m_n}(t) \exp 2\pi i \ (m_1 x_{10} + \ldots + m_n x_{n0})$$

and using the measure-preserving property it is easy to show $\alpha_j(t) = a_j t$. So $x_j(t;\underline{x}_0) = x_{j0} + a_j t + \phi_j(t;\underline{x}_0)$, say. Define $f : R^n \to R^n$ by $f(x_{10}, \ldots, x_{n0}) = (\xi_1, \ldots, \xi_n)$ where

$$\xi_j = x_{j0} + \lim_{T\to\infty} \frac{1}{T} \int_0^T \phi_j(t;\underline{x}_0)\, dt,$$

the limit existing since ϕ_j can be shown to be almost periodic using the stability assumption (2). It is easily verified that f defines a one-valued function on T^n, takes orbits to straight lines, and f is continuous since the limit is uniform. The surjectivity follows from assumption (1); the argument for injectivity is less immediate, using topological group theory to show if $G = \bigcup_{-\infty<t<\infty} \underline{x}(t;\underline{x}_0)$ then (i) the flow structure makes $\bar{G}$ a topological group (by (2)), (ii) $\bar{G} = T^n$ as sets, (ii) $f : \bar{G} \to T^n$ is a group monomorphism.

With the same assumptions, it can also be proved that the flow is ergodic. Further details of the above may be found in [2].

REFERENCES:

[1] Saito, T. On the measure-preserving flow on the torus, J. Math. Soc. Japan 3 (1951) 279-284.

[2] Saito, T. On dynamical systems in n-dimensional torus, Funkcial. Ekvac. 7 (1965) 91-102.

(12) Continuous flows on the plane : techniques I A. Beck

This and the following article give an outline of the techniques used in the proofs of parts (i)', (ii)' of the main theorem of (6) (page 12 above). We consider first flows on $E^2 \cup \{\infty\}$ with no stagnation points at all.

The principal way of studying a flow ϕ in E^2 is to investigate the orbits near a periodic orbit. Let p(x) denote the period of x, i.e. $p(x) = \inf\{t \mid t > 0, \phi(x, t) = x\}$. Then $p : X \to R^+$ is lower semi-continuous for flows on any space X.

When $X = E^2$ we have:

Lemma. If $x_i \to x$ and $p(x) = \infty$ then $p(x_i) = \infty$ for almost all i.

Corollary. $\{x \mid 0 < p(x) < \infty\}$ is closed in $E^2 \setminus F(\phi)$. (Note $F(\phi) = \{x \mid p(x) = 0\}$.)

Let y be a point with $p(y) < \infty$, $y \notin F(\phi)$, and suppose (Case 1) $y \in \omega(x)$ for some x, where as usual $\omega(x)$ denotes $\{w \mid \phi(x,t_k) \to w$ for some sequence $t_k \to +\infty\}$. Then it is possible to show:

Lemma. $y \in \omega(x) \Rightarrow y \in \omega(z)$ for all z sufficiently near the orbit $\theta(y)$ of y and on the same side of $\theta(y)$ as x.

Lemma. $y \in \omega(x) \Longrightarrow \omega(x) = \theta(y), = \omega(z)$ for all z near $\theta(y)$ as above.

Corollary. $y \in \omega(x) \Longrightarrow x \notin \omega(z)$ for any z.

Clearly, similar results hold for α-sets (defined by $t_k \to -\infty$).

The only other possibility is Case 2 : $p(x) < \infty$ for x arbitrarily close to $\theta(y)$. Then it is easy to show:

Lemma. For x close enough to $\theta(y)$ and with $p(x) < \infty$ the orbit $\theta(x)$ is concentric with $\theta(y)$, and all such $\theta(x)$'s are concentric with each other.

Using the above results the configuration of orbits of ϕ can now be pieced together. Start with a periodic orbit $\theta(y)$. In Case 1 it follows that $\{z \mid \alpha(z)$ or $\omega(z) = \theta(y)\}$ is (topologically) an open annulus around $\theta(y)$, and its boundary must lie in $F(\phi)$ or contain a periodic orbit. Thus in Cases 1 or 2 we can build out from $\theta(y)$ producing concentric periodic orbits until (after a countable number of steps) we can go no further: then the set of points on or between these orbits is an open annulus with boundary in $F(\phi)$. Between two 'adjacent' periodic orbits the orbits are spirals of one of two types:

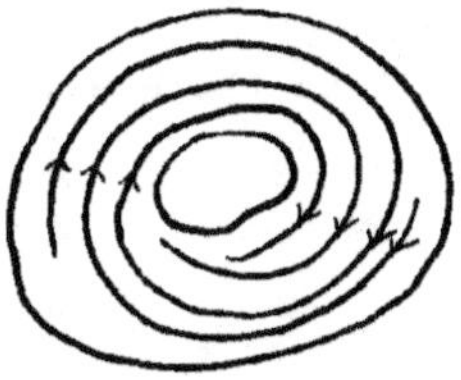

(a)

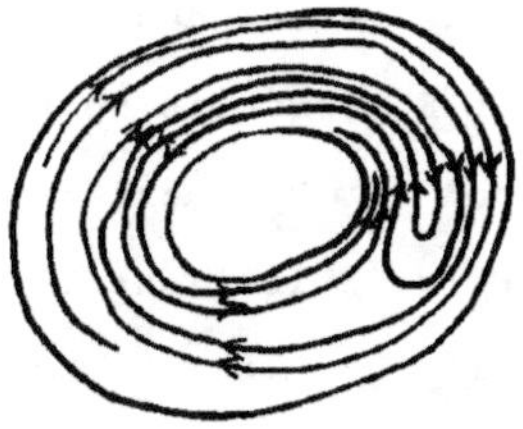

(b)

Between any two periodic orbits there can be only finitely many spirals of type (b), by continuity of ϕ.

<u>THEOREM</u>. Let $B \subset A$ where A is an open annulus and B is a union of concentric Jordan curves each linking the boundary of A. Let B be closed and let C_r be an empty, finite or countable collection of the component annuli of $A \setminus B$ with only finitely many meeting any compact subset of A. Then <u>there is a flow ϕ</u> on $\bar{A}$ with $p(x) < \infty \iff x \in B$, and with a type (b) spiral in a component C of $A \setminus B \iff C \in C_r$ and a type (a) spiral otherwise. Furthermore, for every ϕ and every component A of $E^2 \setminus F(\phi)$ the orbits of ϕ can be represented in this way.

(13) <u>Continuous flows on the plane : techniques II</u> <u>A. Beck</u>

We now investigate flows ϕ on $E^2 \cup \{\infty\}$ with <u>finitely many stagnation points</u>. Let $F_0 = \bigcup$ (components of $F = F(\phi)$ in $E^2 \cup \{\infty\}$ containing stagnation points).

Let $G_0 = G_1 \cup G_2 \cup G_3$ where

$G_1 = \bigcup\{\theta(x) \mid \theta(x)$ has a stagnation point$\}$

$G_2 = \bigcup\{\theta(x) \mid 0 < p(x) < \infty$ and every neighbourhood of $\theta(x)$ meets $G_1\}$

$G_3 = \bigcup\{x \mid p(x) = \infty$ and $(\alpha(x) \cup \omega(x)) \cap (G_1 \cup G_2) \neq \emptyset\}$.

Using the techniques from the previous article it is easy to show that <u>$F \cup G_0$ is closed.</u>

Let F_x = component of x in F ($\subset E^2 \cup \{\infty\}$). After several technical lemmas we can deduce the following:

THEOREM 1. If F_x contains no stagnation point then there exists a neighbourhood $N \supset F_x$ such that $N \cap G_0 = \emptyset$.

Corollary. $F_0 \cup G_0$ is closed.

Proof of Corollary. If $\{x_i\} \subset G_0$ and $x_i \to x \in F$ then F_x contains a stagnation point (by the theorem) so $x \in F_0$. Thus $\bar{G}_0 \cap F = F_0$. But $\bar{G} \subset F \cup G_0$ since $F \cup G_0$ closed, so $\bar{G}_0 \subset (\bar{G}_0 \cap F) \cup G_0 \subset F_0 \cup G_0$.

Every component C of $E^2 \setminus (F_0 \cup G_0)$ is open and thus is a disc or multiple annulus, and only a finite number of such C's are multiple annuli since each divides the set of stagnation points. Now (i) $C \setminus F$ is open, and consists of a countable union of open annuli (see previous article). Also, (ii) every component of $C \cap F$ is compact. We take these two conditions to define a T-set: thus F is a T-set of C. Hence we have:

THEOREM 2. $E^2 \cup \{\infty\} = F_0 \cup \bigcup_i C_i$ where $\{C_i\}$ = countable collection of open sets, finitely many being multiple annuli, and $C_i \cap G_0 = \emptyset \Longrightarrow F \cap C_i$ is a T-set of C_i.

If $C_i \cap G_0 \neq \emptyset$ we at least know the following: $C_i = (G_0 \cap C_i) \cup \bigcup_j C_{ij}$ where each C_{ij} is a disc or a multiple annulus, only finitely many are multiple annuli, $F \cap C_i$ is closed in C_i, and $F \cap C_{ij}$ is a T-set of C_{ij}. We call such a set $F \cap C_i$ a T_σ-set of C_i.

Therefore we have this theorem:

THEOREM 3. If $F = F(\phi)$ with F_0 defined as before then on each component C_i of $E^2 \setminus F_0$ we have

(1) If (a) the boundary ∂C_i of $C_i \ni$ no stagnation point

or (b) The stagnation points of ∂C_i are not accessible from C_i

<u>or</u> (c) the stagnation points of ∂C_i are not accessible from $C_i \setminus (C_i \cap F)$

then (since (a) - (c) $\Longrightarrow C_i \cap G_0 = \emptyset$) $C_i \cap F$ is a T-set of C_i, while

(2) If ∂C_i has stagnation points at least one of which is accessible from $C_i \setminus (C_i \cap F)$ then $C_i \cap F$ is a T_σ-set of C_i.

If there is only one stagnation point then these conditions are necessary and sufficient for F to be $F(\phi)$ for some ϕ; if there are more then deeper analysis is needed.

(14) <u>Generalisation of Bendixson's theory</u> <u>T. Saito</u>

Some well-known properties of the behaviour of orbits of dynamical systems on the plane or 2-sphere (Bendixson; see [1]) cannot be directly generalised to other systems since they depend heavily on 2-dimensional topology, e.g. the Jordan Curve Theorem. However, if we regard Bendixson's work as (1) investigation of the topological behaviour of orbits near a critical point or periodic orbit, and (2) classification of critical points etc. according to their behaviour, this abstract approach can be generalised as follows:

(1) investigation of topological behaviour of orbits near a certain type of subset,

(2) classification of these types according to the above behaviour.

Let $\pi : X \times R \to X$ be a flow on a topological space X; let $C^+(x)$ ($(C^-(X))$) denote the positive (negative) semi-orbit from x, $C(x) = C^+(x) \cup C^-(x)$, and let $L^+(x)$, $L^-(x)$ be the ω- and α-limits of x respectively.

<u>THEOREM 1</u> (Bendixson). If $X = R^2$ (or S^2) and x is an isolated

critical point then either (i) in any neighbourhood of x there are infinitely many periodic orbits surrounding x, or (ii) there is $y \neq x$ such that $L^+(y) = x$ or $L^-(y) = x$.

THEOREM 2 (Bendixson). If $X = R^2$ (or S^2), and C is a periodic orbit then either (i) in any neighbourhood of C there are infinitely many periodic orbits, or (ii) there is $y \in C$ such that $L^+(y) = C$ or $L^-(y) = C$.

If $\mathfrak{M}$ is a family of sets say $M \subset X$ is isolated from $\mathfrak{M}$ if there is a neighbourhood U of M with $N \in \mathfrak{M}$, $N \subset U \Longrightarrow N \subset M$. The following generalises Theorems 1 and 2:

Proposition. If $\mathfrak{M}_0$ = all compact invariant sets of a dynamical system then either (i) M is not isolated from $\mathfrak{M}_0$ or (ii) there is $y \notin M$ with $L^+(y) = M$ or $L^-(y) = M$.

This is known to be true when X is a locally compact metric space, [4].

$M \in \mathfrak{M}_0$ is positively stable if for any neighbourhood $U \supset M$ $\exists$ a neighbourhood $V \subset U$ such that $x \in V \Longrightarrow C^+(x) \subset U$.

Zubov's criterion (which is false) [5] states that M is positively stable $\Longleftrightarrow L^-(X \setminus M) \cap M = \emptyset$ (where $L^-(N)$ means $\bigcup_{x \in N} L^-(x)$): the system $\dot{x} = 0$, $\dot{y} = y^4 \sin\frac{2\pi}{y} + x^2$ is a counterexample in R^2. However, this can be corrected as follows: denote by $D^+(x)$ the first positive prolongation of $x = \{y \mid \exists \{x_n\} \to x, t_n (\geq 0) \to \alpha \leq \infty$, such that $\lim \pi(x_n, t_n) = y\}$.

THEOREM 3 (Ura, [2]). $M \in \mathfrak{M}_0$ is positively stable $\Longleftrightarrow D^+(M) = M$, or equivalently $D^-(X \setminus M) \cap M = \emptyset$ (where D^- is defined by changing the sign of t in the definition of D^+).

Note. D^+ can be generalised to D^+_β for every ordinal β, and then higher orders of stability can be defined by $M = D^+_\beta(M)$, [3].

REFERENCES:

[1] Bendixson, T. Sur les courbes définies par des équations différentielles, Acta Math. 24 (1901) 1-88.

[2] Ura, T. On the flow outside a closed invariant set; stability, relative stability and saddle sets, Contr. Diff. Eqns. III (1964) 249-294.

[3] Ura, T. Sur le courant extérieur à une région invariante, prolongements d'une caractéristique et l'ordre de stabilité, Funkcial. Ekvac. 2 (1959) 143-200.

[4] Ura, T. and I. Kimura. Sur le courant extérieur à une région invariante; théorème de Bendixson, Comm. Math. Univ. Sancti Pauli 8 (1960) 23-39.

[5] Zubov, V.I. The methods of A. Liapunov and their applications, Izdatel'stvo Leningradskogo Univ., Moscow, 1957. English translation AEC-tr. 4439.

(15) Geodesic flows

L.W. Green

The geodesic flow ϕ on the unit tangent bundle V of a (compact) Riemannian manifold M is defined by $\phi_t(v) = (m(t), m'(t))$ where $v = (m_0, e)$ and $m : R \to M$ is the geodesic with initial conditions (m_0, e) and arc-length t as parameter. Much is known about the ergodic properties of ϕ_t [1], but we give here a new method of proving a result in two dimensions using the idea of group representations and having powerful general implications.

THEOREM. If M is 2-dimensional compact oriented, with negative curvature, then ϕ is weakly mixing.

Proof.

Case I: constant curvature. The group $SL(2;R)/Z_2$ acts differentiably on the upper half $\mathbb{C}'$ of the complex plane as a group of Möbius transformations, and induces a transitive action on the unit tangent bundle of $\mathbb{C}'$. As $\mathbb{C}'$ (with the Poincaré metric) is the universal cover of M we have $M = \mathbb{C}'/\pi_1(M)$ so V can be represented in the form $SL(2;R)/\Gamma$. The geodesic flow is then in fact given by $\phi_t(\Gamma g) = \Gamma g g_t$ where

$$g_t = \begin{pmatrix} e^{t/\sqrt{2}} & 0 \\ 0 & e^{-t/\sqrt{2}} \end{pmatrix}.$$

Now we define two more flows given by $\Gamma g \to \Gamma g h_s^{\pm}$ where

$$h_s^+ = \begin{pmatrix} 1 & s \\ 0 & 1 \end{pmatrix}, \quad h_s^- = \begin{pmatrix} 1 & 0 \\ s & 1 \end{pmatrix}.$$

Then $h_s^+ g_t = g_t h^+_{se^{-\sqrt{2t}}}$ and $g_t h_s^- = h^-_{se^{-\sqrt{2t}}} g_t$, so

$$U_{g_{-t}} H_s^+ U_{g_t} = H^+_{se^{-\sqrt{2t}}}, \quad U_{g_t} H_s^- U_{g_{-t}} = H^-_{se^{-\sqrt{2t}}}$$ where U_{g_t}, $H_s^{\pm}$

are the unitary operators on $\mathcal{L}_2(V)$ defined by the measure-preserving transformations ϕ_t, $h_s^{\pm}$ respectively. Since

$H^{\pm}_{se^{-\sqrt{2t}}} \to I$ strongly as $t \to \infty$ for fixed s it follows that if $U_{g_t} f = e^{i\lambda t} f$ then

$$||U_{g_{-t}} H_s^+ U_{g_t} f - f|| = ||H_s^+ U_{g_t} f - U_{g_t} f|| = ||H_s^+ f - f|| = 0$$

since $H_s^+ f - f$ is independent of t and the first term $\to 0$ as $t \to \infty$, and similarly for H_s^-. Hence, the eigenfunction f is invariant under H_s^+ and H_s^-, so since h_s^+, h_s^- together generate $SL(2;\mathbb{R})$ it must be constant a.e.. Thus ϕ_t is weakly mixing.

Case II: non-constant curvature. The relation $h_s^+ g_t = g_t h^+_{se^{-\sqrt{2t}}}$ above results in fact from the identity $[X,H^+] = \sqrt{2}H^+$, where X, H^+ are the infinitesimal generators of the flows ϕ_t, h_s^+. In general, we have the following:

Lemma. Let H,X (with associated flows H_s, X_t considered also as operators in $\mathcal{L}_2(V)$) be vector fields on V of class $C^p (p \geq 2)$ such that $[X,H] = \lambda H$ where λ is a positive continuous function on V. If X_t is measure-preserving then $X_{-t}H_sX_t \to I$ strongly as $t \to \infty$ for each s.

From this it follows as in Case I that <u>any eigenfunction of the flow X_t is invariant under H_s</u>. Therefore, we seek (two) solutions H of $[X,H] = \lambda H$ which will in some sense 'generate' the tangent space of V, where $X_t = \phi_t$. Associated with X on V is the vector field Y of the 'perpendicular' geodesic flow, defined by taking geodesics initially perpendicular to those defining ϕ_t (remembering M is oriented); let A denote the field of 'rotation around the fibres of V'. Then we have

$$[X,A] = -Y \quad , \quad [Y,A] = X \quad , \quad [X,Y] = kA$$

where k = Gaussian curvature. Trying $H = Y + uA$ we find $[X,H] = \lambda H$ $(\lambda = -u)$ if u satisfies the Ricatti equation $X(u) + u^2 + k = 0$. Now this has two solutions $u^{\pm}$, bounded below and above zero respectively, whose geometric interpretation is that $u^{\pm}(v)$ is the geodesic curvature of an orthogonal trajectory of geodesics positively (negatively) asymptotic to the geodesic with initial conditions v [2]. Therefore, writing $H^{\pm} = Y + u^{\pm}A$ we have $H_s^{\pm}f = f$ for any eigenfunction f of X_t, so $H^{\pm}f = 0$. A rough argument then says $(u^+ - u^-)Af = 0$ since $H^+ - H^- = (u^+ - u^-)A$, so $Af = 0$ since $u^+ - u^-$ is bounded away from zero. Hence $Yf = 0$, and $Xf = 0$ since $X = [Y,A]$. Strictly, close attention must be paid to the domains of these operators and more work must be done, since one must justify applying e.g. $H^+ - H^- = (u^+ - u^-)A$ to a function f which is not a priori smooth. However, it can be shown that $Af = Yf = Xf = 0$ in any case, and so f = constant a.e., which again proves the result.

In n dimensions analogues of X,Y,A can be defined and a similar theorem proved, but it has not so far been possible to extend enough of the analysis to give a proof of the theorem in the case of the principal bundle.

REFERENCES:

[1] Anosov, D.V. and J.G. Sinai. Some smooth ergodic systems, Russian Mathematical Surveys 22 (1967) 103-167.

[2] Hopf, E. Statistik der Lösungen geodätischer Probleme von unstabilen Typus, II, Math. Ann. 117 (1940) 590-608.

(16) Instability

M. Shub

Diffeomorphisms $f : X \to X$, $g : Y \to Y$ are topologically conjugate if there is a homeomorphism $h : X \to Y$ such that $hf = gh$. Let M be a C^∞ manifold with $\partial M = \emptyset$, and $\mathrm{Diff}^r(M)$ the space of C^r diffeomorphisms of M, with the C^r topology: then $f \in \mathrm{Diff}^r(M)$ is structurally stable if there is a neighbourhood U_f of f in $\mathrm{Diff}^r(M)$ such that $g \in U_f \Longrightarrow f,g$ topologically conjugate. It was conjectured but proved to be false [2] that structurally stable diffeomorphisms were always dense in $\mathrm{Diff}^r(M)$.

Define $\Omega(f) = \{x \mid$ given open $U \ni x$ there exists $m > 0$ with $f^m(U) \cap U \neq \emptyset\}$. Then $f \in \mathrm{Diff}^r(M)$ is Ω-stable if $\exists\, U_f$ with $g \in U_f \Longrightarrow f|\Omega(f), g|\Omega(g)$ topologically conjugate. An example due to Abraham and Smale in 1967 [1] showed that Ω-stable functions also fail to form an open dense set in $\mathrm{Diff}^r(M)$. However, we may define a weaker property: $f \in \mathrm{Diff}^r(M)$ is topologically Ω-stable if $\exists\, U_f$ with $g \in U_f \Rightarrow \Omega(f)$ homeomorphic to $\Omega(g)$.

Problem: Do topologically Ω-stable diffeomorphisms form an open dense set in $\mathrm{Diff}^r(M)$?

The following shows that toplogical Ωstability is weaker than Ω-stability:

THEOREM. There exists an open set $U \subset \mathrm{Diff}\,(T^2 \times T^2)$ with $g \in U \Longrightarrow g$ topologically Ω-stable but not Ω-stable.

The set U is constructed using an Anosov diffeomorphism $A : T^2 \to T^2$ and a related 'deformed' diffeomorphism DA defined by splitting a saddle of A into a sink and two saddles by a large C^1-isotopy. A diffeomorphism $G : T^2 \times T^2 \to T^2 \times T^2$ is defined by $G(x,y) = (A^2x, \phi(x)y)$ where $\phi : T^2 \to \mathrm{Diff}(T^2)$ is constructed so that $\phi(x) = A$ outside a small disc of T^2, $\phi(x) = DA$ inside a smaller disc, and ϕ gives the isotopy $A \simeq DA$ in between. Then $\Omega(G) = T^2 \times T^2$, which is $\Omega(G')$ for all G' close enough to G by the "equivariant fibration theorem" (p. 40). But G is not structurally stable and so not Ω-stable.

REFERENCES:

[1] Abraham, R. and S. Smale. (To appear in Proceedings of AMS Summer Institute on Global Analysis, Berkeley 1968.)

[2] Smale, S. Structurally stable systems are not dense, Amer. J. Math. 88 (1966) 491-496.

(17) One-parameter families of diffeomorphisms P.Brunovký

We give some results concerning the generic behaviour of periodic points of C^r maps $(1 < r < \infty)$ $f : P \times M \to M$, where P and M are C^r manifolds, dim P = 1, and the map $F_\mu : M \to M$ given by $f_\mu(x) = f(\mu,x)$ is a diffeomorphism for each $\mu \in P$. For given P,M we denote by $\mathfrak{F}$ the set of all such f's with the Whitney C^r topology.

Similar problems have been studied by J. Sotomayor (two-dimensional flows) in [2] and K. Meyer (two-dimensional symplectic diffeomorphisms) in [1].

Denote by $Z_k \subset P \times M$ the set of all k-periodic points of f, i.e. $Z_k = \{(\mu,x) \mid f_\mu^k(x) = x,\ f_\mu^j(x) \neq x \text{ for } 0 < j < k\}$.

THEOREM 1. For every f belonging to a residual subset $\mathfrak{F}_1 \subset \mathfrak{F}$:

(a) Z_k are 1-dimensional submanifolds of $P \times M$; Z_1 is embedded and closed;

(b) for fixed μ and k, the k-periodic points of f_μ are isolated;

(c) the set Y_k of all $(\mu,x) \in Z_k$ such that $df_\mu^k(x) - \text{id.}$ is singular, is discrete; for any $(\mu_o,x_o) \in Z_k \setminus Y_k$ there is a neighbourhood $W = U \times V$ of (μ_o,x_o) and a C^r map $\phi : U \to V$ such that $Z_k \cap W$ is equal to the graph of ϕ.

The following theorem examines in more detail the behaviour of f in the vicinity of a point of Y_k.

THEOREM 2. For every f belonging to a residual subset $\mathfrak{F}_2 \subset \mathfrak{F}_1 \subset \mathfrak{F}$ and any $(\mu_o,x_o) \in Y_k$:

(a) one of the eigenvalues of $df_{\mu_o}^k(x_o)$ is +1, the others are not on the unit circle ;

(b) there is a neighbourhood $W = U \times V$ of (μ_o,x_o) such that every point $(\mu,x) \in (Z_k \cap W) \setminus \{(\mu_o,z_o)\}$ satisfies $\mu < \mu_o$ (P being locally R);

(c) $Z_k \cap W$ is connected and the number of eigenvalues of $df_\mu^k(x)$ outside the unit circle at points of different components of $(Z_k \cap W) \setminus \{(\mu_o,x_o)\}$ differs by one ;

(d) $W \setminus Z_k$ contains no invariant set of f^k.

Remark: If we restrict ourselves in Theorems 1 and 2 to the case k = 1, residual can be replaced by open dense.

In virtue of Theorem 2, we call the points of Y_k collapsation points. Namely, for $\mu < \mu_o$, $\mu \in U$, $Z_k \cap (\{\mu\} \times V)$ consists of two points that collapse at $\mu = \mu_o$ and disappear for $\mu > \mu_o$.

For $\dim M = 2$, the behaviour of f^k in the neighbourhood of a collapsation point can be illustrated by the following picture (up to inversion of f):

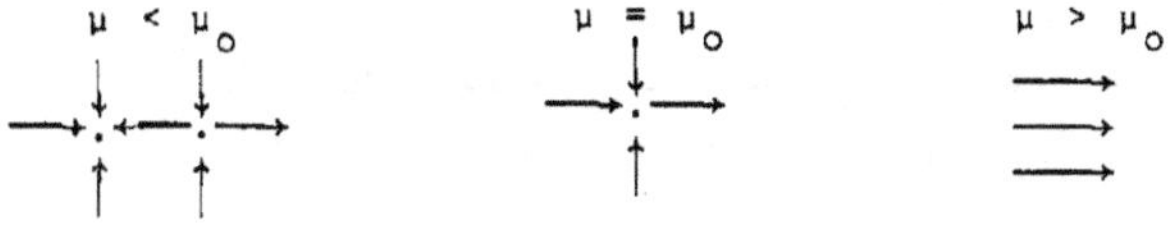

Besides the sets Z_k it is interesting also to study the sets $\bar{Z}_k$ and the behaviour of f in the vicinity of the points of $\bar{Z}_k \setminus Z_k$. Namely, such points are, for continuity reasons, ℓ-periodic points with ℓ being a divisor of k. We call the points of $\bar{Z}_k \setminus Z_k$ <u>bifurcation</u> points.

A k-periodic point (μ,x) can be a bifurcation point only if $df_\mu^k(x)$ has some root of 1 as eigenvalue. Therefore we need some information on how the eigenvalues cross the unit circle S in the complex plane if μ is changed.

The results which follow are true for $\dim M = 2$, although it can be expected that similar theorems are true for higher dimensions.

Denote $D_k = \{(\mu,x) \in Z_k \mid df_\mu^k(x) \text{ has double roots}\}$. The eigenvalues $\lambda_1^{(k)}$, $\lambda_2^{(k)}$ of $df_\mu^k(x)$ are C^r functions on $Z_k \setminus D_k$.

<u>THEOREM 3</u>. For every f in a residual subset $\mathfrak{F}_3 \subset \mathfrak{F}_2 \subset \mathfrak{F}$, the following is true:

(a) $\lambda_i^{(k)}(D_k) \cap S = \emptyset$, $i = 1,2$;

(b) $\lambda_1^{(k)} \pitchfork S$, $\lambda_2^{(k)} \pitchfork S$;

(c) If, for some $(\mu,x) \in Z_k$, $\lambda_1^{(k)}(\mu,x) \in S$, then either $\lambda_2^{(k)}(\mu,x) \notin S$ or $\lambda_1^{(k)}(\mu,x)$ is not a root of 1.

From Theorems 2,3 it follows that, generically, (μ,x) can be a bifurcation point only if one of the eigenvalues of

$df^k_\mu(x)$ is -1, the other being real $\neq \pm 1$. We denote by X_k the set of such points.

THEOREM 4. Assume $r \geq 3$. Then, for every f belonging to a residual subset $\mathfrak{F}_4 \subset \mathfrak{F}_3 \subset \mathfrak{F}$:

(a) $(\mu_o, x_o) \in Z_k$ is a bifurcation point iff $(\mu_o, x_o) \in X_k$;

(b) If $(\mu_o, x_o) \in X_k$, then there is a neighbourhood W of (μ_o, x_o) such that $(Z_{2k} \cap W) \setminus \{(\mu_o, x_o)\}$ consists of two components separated by (μ_o, x_o), and every point $(\mu, x) \in (Z_{2k} \cap W) \setminus \{(\mu_o, x_o)\}$ satisfies $\mu < \mu_o$;

(c) Either the points of $Z_k \cap W$ are sources for $\mu \geq \mu_o$ (degenerate for $\mu = \mu_o$), saddles for $\mu < \mu_o$ and the points of $Z_k \cap W$ are sources, or the same is true with source replaced by saddle and conversely, or one of the above cases is true for f^{-1}.

(d) $W \setminus (Z_k \cup Z_{2k})$ contains no invariant set of f^k.

The behaviour of f^k in the neighbourhood of a bifurcation point can be illustrated by one of the following pictures (up to inversion of f):

$\mu < \mu_o$ $\mu = \mu_o$ $\mu > \mu_o$

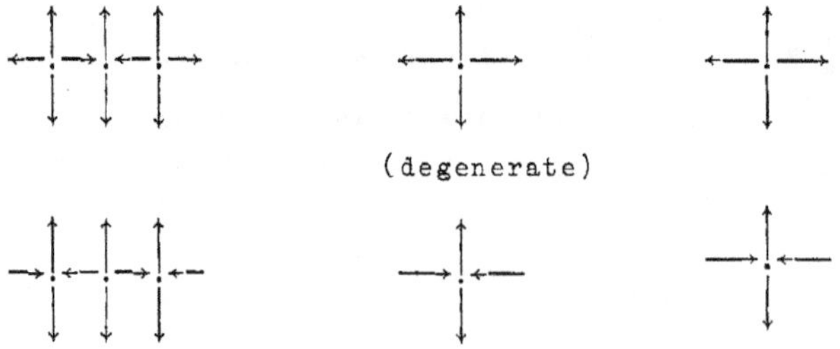

THEOREM 5. For every $f \in \mathfrak{F}_4$:

(a) for k odd, Z_k is a closed embedded submanifold of $P \times M$

(b) for k even, $\bar{Z}_k$ is a closed embedded C^1 submanifold of $P \times M$; $\bar{Z}_k \setminus Z_k$ is discrete and coincides with $X_{k/2}$.

REFERENCES:

[1] Meyer, K. Generic bifurcation of periodic points (to appear).

[2] Sotomayor, J. Generic one-parameter families of vector fields, Bull. A.M.S. 74 (1968) 722-726.

(18) Topology and Mechanics

S. Smale

We consider the 3-body problem: point masses m_1, m_2, m_3 in R^3 are acted on only by their mutual gravitational attraction, and we seek a qualitative description of their motion. If m_i has position $\underline{x}_i \in R^3$ the total configuration is given by a point in $R^9 - \Delta = M^9$, say, where Δ= points with $\underline{x}_i = \underline{x}_j$ (some $i \neq j$) since we assume no collisions. Kinetic energy defines a metric on M^9 by the form $K = \Sigma \frac{1}{2} m_i \, || \dot{\underline{x}}_i ||^2$, and potential energy $V : M^9 \to R$ is given by $\Sigma -m_i m_j / || \underline{x}_i - \underline{x}_j ||$. Then $H = K + V\pi : TM^9 \to R$ (where $\pi : TM^9 \to M^9$ is the tangent bundle) defines a vector field X_H on TM^9 in a standard way (see e.g. [1]).

THEOREM (Lagrange-Jacobi). The order of X_H (i.e. the dimension of the space on which it is defined) can be reduced from 18 to 7.

We give a modernized proof which also explains what 'reduced' means.

Since the centre of mass moves uniformly in R^3 we can restrict attention first of all to an invariant 12-dimensional manifold $TM^6 = \{u \in TM^9 \mid \text{c.m. fixed at } \underline{O}\}$.

If G is a Lie group acting as a group of isometries on a manifold M there is a natural association of a vector field X* on M with every X $\in$ the Lie algebra $\mathcal{G}$ (=tangent space of G at e). In our case $M = M^6$, $G = SO(3)$. For $X \in \mathcal{G}$ define $f_X : TM^6 \to R$ by $f_X(Y_X) = \langle X^*_x, Y_x \rangle_K$ (inner product using K) : then, since V is G-invariant f_X is invariant under the flow of X_H - i.e. an <u>integral</u> for X_H. Three independent integrals (angular momentum) are obtained in this way, and with H they reduce the order of X_H to 12 - 4 = 8.

Roughly speaking, there are no more integrals. However, there is a further reduction called classically "elimination of the mode", corresponding to the fact that, if the initial situation is rotated by ϕ_0 about the angular momentum axis, all future situations are also rotated by ϕ_0. Define $J : TM \to \mathcal{G}^*$ (=dual of $\mathcal{G}$) by $J(X_x, Y) = \langle X_x, Y^+_x \rangle_K$ ($Y \in \mathcal{G}$; Y^*_x as above).

<u>Proposition</u> (valid for any G leaving some K,V invariant in a mechanical problem). J is equivariant with respect to the adjoint G-action on $\mathcal{G}^*$ (defined from derivative of G-action on G given by $\alpha \to \{x \to \alpha x \alpha^{-1}\}$).

<u>Corollary</u>. $G_p = \{g \in G \mid g.p = p\}$ ($p \in \mathcal{G}^*$) leaves $J^{-1}(p)$ invariant, so we get an induced flow on $J^{-1}(p) / G_p$.

If $G = SO(3)$ then $\mathcal{G}^* = R^3$ and $G_p = S^1$ which acts on 8-dimensional $(H \times J)^{-1}(h,p)$ ($h \in R$) giving a quotient space of dimension 7.

The main problem is to analyse the structure of $H \times J$, e.g. by finding the topological structure of $J^{-1}(p) / G_p$. Quite complete results can be obtained for the (much easier) 2-body problem.

<u>REFERENCES</u>:

[1] Abraham, R. and Marsden, J.E. Foundations of Mechanics, <u>Benjamin, 1967.</u>

(19) Generic properties of conservative systems R.C.Robinson

The Kupka-Smale theorem states that for a residual subset of all diffeomorphisms of a compact manifold (i) the periodic points are hyperbolic, and (ii) the stable and unstable manifolds meet transversally.

Part (i) no longer holds if we consider only symplectic diffeomorphisms, so we give a new condition which replaces hyperbolic; however, part (ii) still holds for symplectic diffeomorphisms.

Let M^{2n} be a symplectic manifold with closed non-degenerate 2-form ω. Say $f \in \mathrm{Diff}^r(M)$ is symplectic ($f \in \mathrm{Sym}^r(M)$) if $f^*\omega = \omega$. Let m be a periodic point of period p : then $Tf^p(m)$ is a symplectic transformation with eigenvalue λ_i, say. These occur in conjugate and inverse pairs, so we define the principal eigenvalues to be half of those of value +1 or -1, together with those satisfying (1) $|\lambda_i| > 1$ or (2) $|\lambda_i| = 1$, $\mathrm{Im}\,\lambda_i > 0$. If $N \in Z$, say m is N-generic if the principal eigenvalues are multiplicatively independent over $[-N,N] \cap Z$ ($\prod_i \lambda_i^{n_i} = 1$, $n_i \in [-N,N] \cap Z \Rightarrow n_i = 0$).

THEOREM. (i) $\mathcal{E}^r(p,N) = \{f \in \mathrm{Sym}^r(M) \mid$ points of period $\leq p$ are N-generic$\}$ is open and dense in $\mathrm{Sym}^r(M)(1 \leq r \leq \infty)$; $\mathcal{E}^r(\infty,\infty)$ is residual.

(ii) for a residual subset of $\mathrm{Sym}^r(M)$ the stable and unstable manifolds meet transversally.

In dimension 2 with $r \geq 4$ it can be shown that generically all elliptic ($|\lambda_i| = 1$) periodic points are of stable type. Looking at such a point it can be shown that there is in $\mathrm{Sym}^r(M)$ an open set of non-Ω-stable diffeomorphisms.

We now turn to flows. A C^r function $H : M \to R$ induces a 'Hamiltonian' vector field X_H on M and a flow ϕ in a standard way. If m is a point with finite period τ then $T_m\phi(\tau,m)$ has 2n eigenvalues of which two are +1. From the other 2n-2 define principal eigenvalues as above, and define the orbit of m to be N-generic as above and similarly for critical points. Let $\mathcal{G}^r(p,N)$ be the set of Hamiltonian vector fields satisfying (1) all critical points are N-generic, (2) all closed orbits lie on 1-parameter families of closed orbits, (3) all but a countable number of closed orbits of period $\leq$ p are N-generic, and (4) the non-N-generic closed orbits satisfy a transversality condition.

THEOREM. $\mathcal{G}^r(p,N)$ is residual ($N \leq \infty$) in the space of Hamiltonian fields.

The types of matrices with multiple eigenvalues that occur generically can be given explicitly.

For full details of all the above, see [1].

REFERENCES:

[1] Robinson, R.C. Global properties of Hamiltonian systems, (thesis, Berkeley 1969. To appear in Proceedings of AMS Summer Institute on Global Analysis, Berkeley 1968).

(20) Dynamical systems on nilmanifolds — W. Parry

If G is a locally compact group of continuous measure-preserving transformations of a compact space X with normalized measure, then G ergodic means $fg = f$ ($f \in \mathcal{L}_2$, all $g \in G$) $\Rightarrow$ f = constant a.e., and G weakly mixing means $fg = \alpha(g)f$ ($\alpha \in \hat{G}$) $\Rightarrow$ f constant a.e.. If G,H are two groups as above with (1) $gh = \lambda_g(h)g$ ($\lambda: G \to \mathrm{Aut}(H)$) and (2) $\overline{\{\lambda_g(h) \mid g \in G\}} \ni$ identity of H (all $h \in H$) then it can easily be shown that any $f \in \mathcal{L}_2$ satisfying $f(gx) = \alpha(g)\, f(x)$ ($\alpha \in \hat{G}$, all $g \in G$) is H-invariant ('Mauntner phenomenon', cf. lecture (15) by L. Green (p.25)). In particular,

H ergodic $\Rightarrow$ G weakly mixing.

Let D be a uniform discrete subgroup of a simply-connected nilpotent Lie group N (ie. N/D compact), and give N/D the unique (normalized) measure invariant under the action of N. Let $A : N \to N$ be an automorphism with $A(D) = D$ so inducing A on the nilmanifold N/D. Say A is <u>ergodic</u> if $Z = \{A^n\}$ is ergodic.

<u>THEOREM</u>. Let $N^1 \doteq [N,N]$. Then A is ergodic (or weakly mixing) on N/D $\Leftrightarrow$ A is ergodic on $N/N^1.D$.

<u>Proof</u>. $\Rightarrow$ is obvious, so we prove $\Leftarrow$. Let $H = \{g \in N | A^n g \to e\}$; take $G = Z$ acting by $nxD = A^n(x)D$. Now $A^n(hxD) = A^n(h)A^n(x)D$, i.e. $nh = \lambda_n(h).n$ where $\lambda_n(h) = A^n(h)$. Also $\overline{\{A^n(h) | n \in Z\}} \ni$ identity of H, so the above applies and it thus suffices to prove H contains a 1-parameter ergodic group. A theorem of Green says that a 1-parameter group on N/D which projects to an ergodic 1-p.g. on $N/N^1.D$ is itself ergodic. Since H projects <u>onto</u> $H_1 = \{g \in N/N^1 | A^n g \to e\}$ (proof by linear algebra) and there is essentially a unique 1-p.g. containing any $g \neq e \in N$, we are reduced to finding an ergodic 1-p.g. contained in H_1. Now $N/N^1 = R^n$ and $N/N^1.D$ must be a torus $K^m = R^m/Z^m$. Let $H_2 = \{h \in K^m | A^n h \to e\}$. Clearly $A(\bar{H}_2) = \bar{H}_2$, and $A : K^m/\bar{H}_2 \to K^m/\bar{H}_2$ is ergodic with eigenvalues on the unit circle. These have to be roots of unity, which contradicts ergodicity. So $\bar{H}_2 = K^m$. Now H_1 contains w with rationally independent coordinates (otherwise there exists $N \neq \underline{0} \in Z^m$ with $\langle N,v\rangle = 0$, all $v \in H_1$, which defines a character annihilating H_2 and thus K^m) and w determines an ergodic 1-p.g. in H_1.

The following can also be proved:

(i) A weakly mixing $\Rightarrow$ A is a K-automorphism;

(ii) Affine $T : N/D \to N/D$ is minimal (in fact uniquely ergodic)

$\Leftrightarrow$ induced $T' : N/N^1.D \to N/N^1.D$ is minimal

$\Leftrightarrow$ T is ergodic with zero entropy

$\Leftrightarrow$ T is ergodic and distal.

(21) Linearizing a diffeomorphism C. Pugh

Let $f:M \to M$ be a C^1 diffeomorphism of a compact C^∞ manifold M ($\partial M = \emptyset$), with V an invariant C^1 submanifold of M ($f(V) = V$). Say f is normally hyperbolic at V if $TM|V = TV \oplus N^u \oplus N^s$ where TV, N^u, N^s are invariant under Tf and $||N^s f|| < m(Tf|TV)$, $m(N^u f) > ||Tf|TV||$. (Here $m(L)$ means $||L^{-1}||^{-1}$, $N^s f = Tf|N^s$ etc. and we assume some Riemannian metric.) Thus "behaviour in the $N(=N^u \oplus N^s)$ direction dominates that in TV".

THEOREM. f normally hyperbolic $\Rightarrow$ f conjugate to Nf near V. (When V is a point this is Hartman's theorem. For V a closed orbit of a flow the result was proved by Shub and Irwin).

Proof. By standard theory there are C^1 stable and unstable manifolds $W^s = \{x \in M |\ f^n(x) \to V,\ n \to \infty\}$, $W^u = \{x \in M | f^{-n}(x) \to V, n \to \infty\}$ with W^s tangent to $TV \oplus N^s$, W^u to $TV \oplus N^u$. For $p \in V$ define $W^{ss}(p) = \{x \in M | d(f^n x, f^n p) \leq k^n d(x,p)\}$ where $||N^s f|| \leq k < m(Tf|TV)$. Again by stable manifold theory this gives locally an f-invariant fibration of $W^s(V)$ (with C^1 fibres varying continuously) and a local homeomorphism $\gamma : N^s_\varepsilon \to W^s$ with $\gamma(N^s_\varepsilon(p)) = W^{ss}(p)$, where N^s_ε is a neighbourhood of V in N^s; $\gamma|V$ = identity, γ is uniformly C^1 along the fibres of N^s_ε, and $(D\gamma)_p$ = identity ($p \in V$). Now along fibres $N^s f$ is C^1-close to $\hat{f} = \gamma^{-1} f \gamma : N^s_\varepsilon \to N^s_\varepsilon$, so since they are both fibre maps covering f on V they are conjugate (by an easy extension of Hartman's theorem). If $h.N^s f = fh$ (say) we have $h_s . N^s f = f.h_s$ where $h_s = \gamma h : N^s_\varepsilon \to W_s$.

Next define a similar fibration W^{uu} of W^u by considering f^{-1} instead of f. Using a construction due to Palis, W^{uu} can be extended to an f-invariant fibration $\tilde{W}^{uu}$ over a neighbourhood of V in W^s. If $x \in N^s(p)$ define $N^u(x) = x + N^u(p) \subset N(p)$; then near V we construct diffeomorphisms $g(x) : N^u(x) \to \tilde{W}^{uu}(h_s x)$ such that $g(x)^{-1}.f.g(x)$ is C^1-close to Nf on $N^u(x)$ (for example by using a projection of $N^u(x)$ onto $\exp_p^{-1}\tilde{W}^{uu}(h_s x)$) and then $\{g(x) | x \in N^s(p)\}$ defines g from a neighbourhood of V in N to a neighbourhood of V in M such that $g^{-1}fg$ is conjugate (by the extended Hartman's theorem) to Nf. Hence f is conjugate to Nf.

<u>COROLLARY</u>. If the restriction of f to V is structurally stable, then f is structurally stable in a neighbourhood of V.

(22) <u>Topologically transitive diffeomorphisms of T^4</u> <u>M. Shub</u>

Let $f:M \to M$ be a diffeomorphism (M compact, connected, C^∞) with p a periodic point of f of period n. The following is easy to prove:

<u>THEOREM 1</u>. Let $W^s(p)$ $(W^u(p))$ be the "stable (unstable) manifold" tangent to the subspace of TM(p) corresponding to eigenvalues of $Df^n(p)$ with modulus <1 (>1) (f need not be hyperbolic). If $W^s(p)$ and $W^u(p)$ are dense in M then f is topologically transitive.

<u>Question</u>: Let A be an ergodic automorphism of T^n. Is a $C^1(C^2)$ perturbation of A topologically transitive?

Let E be a compact C^∞ manifold $(\partial E = \emptyset)$, Λ a topological space. A locally trivial fibration $\pi:E \to \Lambda$ is a <u>C^r-regular fibration</u> if $\pi^{-1}(\lambda)$ is a C^r-submanifold of E $(\lambda \in \Lambda)$ and the map $x \mapsto T_x\pi^{-1}(\pi x)$ is continuous. A <u>perturbation</u> of π is a homeomorphism $h:E \to E$, C^r on fibres, such that πh^{-1} is a C^r-regular fibration, h is C^0-close to the identity, and Dh along fibres is C^1-close to the identity. We say π is a C^r-equivariant fibration for a diffeomorphism $F:E \to E$ and homeomorphism $f:\Lambda \to \Lambda$ if $\pi F = f\pi$.

THEOREM 2. Let $f:M \to M$ be an Anosov diffeomorphism with periodic points dense in M, $f(m_0) = m_0$, $\pi:E \to M$ a C^r-equivariant fibration for $F:E \to E$ and f such that $F|\pi^{-1}(m_0)$ is Anosov with periodic points dense. Then F is topologically transitive.

(The proof uses standard stable manifold theory and the fact that $W^s(\pi^{-1}(m_0)) = \pi^{-1}(W^s(m_0))$.)

THEOREM 3. (Equivariant fibration theorem). Let $\pi:E \to M$ be a C^r-equivariant fibration for F, f where f is Anosov. If f is "more hyperbolic than F" along the fibres then for any sufficiently small C^1-perturbation G of F there is a perturbation π' of π such that π' is a C^1-equivariant fibration for G, f.

Using this we obtain

THEOREM 4. Let π, F, f be as in Theorems 2, 3. Then any sufficiently small C^1 perturbation of F is topologically transitive.

As a special case we have the example on T^4 as described in (16) (page 28).

These two lectures represent part of joint work with Hirsch and Pugh, which will appear.

(23) Ω-explosions J. Palis

A diffeomorphism f of a manifold M (M compact) satisfies Axiom A if the non-wandering set $\Omega = \Omega(f)$ has a hyperbolic structure, and $\overline{\text{Per}(f)} = \Omega$. There is then a 'spectral decomposition' of Ω into components Ω_i on each of which f is topologically transitive (see [3]). An n-cycle on Ω is a sequence $\Omega_0, \Omega_1, \ldots, \Omega_{n+1}$ with $W^s(\Omega_i) \cap W^u(\Omega_{i+1}) \neq \emptyset$, $\Omega_{n+1} = \Omega_0$ and otherwise $\Omega_i \neq \Omega_j$ for $i \neq j$.

THEOREM [2]. If f satisfies axiom A and there is an n-cycle on

$\Omega(f)$ then f is not Ω-stable.

Example (Smale). $M = S^2$, Ω finite.

SO = source
S1 = sink

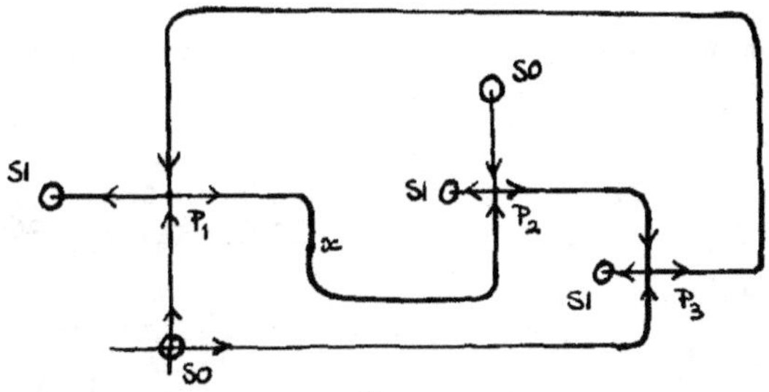

Let U be a small neighbourhood in S^2 of $x \in W^u(P_1) \cap W^s(P_2)$ and $i : S^2 \to S^2$ a diffeomorphism which is the identity outside U, and such that $i(W^u(P_1))$ meets $W^u(P_1)$ transversally inside U. It is easy to verify that if U is small enough then for $i \circ f$ we have $W^u(P_1)$ transversal to $W^s(P_2)$ and $W^u(P_2) \subset \overline{W^u(P_1)}$, $W^s(P_1) \subset \overline{W^s(P_2)}$. Perturbing similarly near a point on $W^u(P_2) \cap W^s(P_3)$, we obtain a situation where the whole of $W^s(P_1) \cap W^u(P_3)$ is non-wandering - so Ω is no longer finite, and in fact infinitely many new periodic points appear.

The proof of the theorem was inspired by this example. As stated, the theorem also holds for flows, although the proof is somewhat different (e.g. $W^u(P_1)$ and $W^s(P_2)$ above cannot meet transversally); for flows Ω also becomes much larger under suitable perturbations - hence the name 'explosion'.

Using [1] we have

THEOREM. If $\Omega(f)$ is finite then f is Ω-stable if and only if $\Omega(f)$ is hyperbolic and there is no cycle on $\Omega(f)$.

The same result holds for flows, when we consider Ω to be the finite union of critical points and closed orbits.

An important result is

THEOREM (Smale [4]). If f satisfies Axiom A and there is no cycle on $\Omega(f)$ then f is Ω-stable.

So there remains the problem: does Ω-stability imply <u>Axiom A</u>?

Besides its natural role in Ω-stability, the cycle property on Ω seems to be relevant in Bifurcation Theory.

REFERENCES:

[1] Palis, J. On Morse Smale dynamical systems, <u>Topology</u> 8 (1969) 385-404.

[2] Palis, J. A note on Ω-stability, (to appear in <u>Proceedings of AMS Summer Institute on Global Analysis, Berkeley 1968</u>).

[3] Smale, S. Differentiable dynamical systems, <u>Bull. AMS</u>. (1967) 747 - 817.

[4] Smale, S. The Ω-stability theorem (to appear in <u>Proceedings of AMS Summer Institute on Global Analysis, Berkeley 1968</u>).

(24) <u>Singularities of exponential maps</u> <u>A. Weinstein</u>

Let TM, T*M be the tangent, cotangent bundles (respectively) of a C^∞ manifold M^n. A Riemannian metric can be considered as a bundle map $\gamma : TM \to T^*M$, giving an 'energy function' $E^\gamma : TM \to R$ by $v \mapsto \frac{1}{2}\langle\gamma(v), v\rangle$. The natural sympletic structure Ω on T*M induces $\gamma^*(\Omega)$ on TM, and hence a vector field X^γ on TM in a standard way. Integrating X^γ (assuming TM complete) gives a flow Exp_t^γ on TM; let $Exp^\gamma = Exp_1^\gamma$, $exp^\gamma = \pi . Exp^\gamma$ (where $\pi : TM \to M$), $exp_p^\gamma = exp^\gamma \mid T_pM$ ($p \in M$). We investigate singularities of exp_p^γ.

For any smooth $f : R^n \to R^n$ we have the jacobian function $j(f) : R^n \to L(R^n, R^n)$ given by the matrix $\frac{\partial f_i}{\partial x_j}$, and determinant $\det : L(R^n, R^n) \to R$. Let $\Sigma_k = \{\phi \in L(R^n, R^n) \mid \text{rank } \phi = n - k\}$; then $\det^{-1}(0) = \bigcup_{0<k\le n} \Sigma_k = \Sigma$, say.

<u>THEOREM 1</u> (Thom). For almost all f, j(f) is transversal to Σ (i.e. transversal to each Σ_k, which is a manifold of

codimension k^2 in $\mathbf{L}(R^n, R^n)$). For such f, $j(f)^{-1}(\Sigma_k)$ is a submanifold of R^n of codimension k^2.

Note that $j(f)^{-1}(\Sigma_k)$ is defined also for $f : M_1^n \to M_2^n$, although j depends on the local coordinates.

THEOREM 2. For all γ there is a smooth function $\hat{j}(\exp_p^\gamma) : R^n = T_pM \to S(R^n)$ (the group of symmetric matrices) such that $\hat{j}^{-1}(\Sigma) = j^{-1}(\Sigma)$ and $\hat{j}$ is transversal to $\Sigma \cap S(R^n)$ for almost all γ.

The way in which symmetric matrices arise can be seen as follows. For manifolds X^n, M^n define a quasi-exponential mapping to be a pair (f, g) where $g : X \to T^*M$ is an immersion which is a lift of $f : X \to M$, and $g^*(\Omega) = 0$. Thus $Tg(x) : T_xX \to T_{g(x)}(T^*M)$ has image V^n such that $\Omega \mid V^n = 0$. We call the set of such V^n the Lagrangian Grassmannian in each fibre of $T(T^*M)$, so $x \mapsto V^n$ defines a 'lifting' $g_\# : X \to L(T^*M)$ of g into the Lagrangian Grassmannian bundle over T^*M. Let ${}_kL(T^*M) \subset L(T^*M)$ denote those V^n whose intersection with the vertical fibre has dimension k. Then it can be shown that $g_\#^{-1}({}_kL(T^*M)) = j(f)^{-1}(\Sigma_k)$. Taking coordinates in each fibre R^{2n} of $T(T^*M)$, it is easy to identify each fibre of ${}_kL(T^*M)$ locally with $\Sigma_k \cap S(R^n)$.

For exponential maps $\exp_p^\gamma$ there is just this situation : g is $\gamma \circ \mathrm{Exp}^\gamma \mid T_pM : T_pM \to T^*M$, and f is $\exp_p^\gamma$; define $\hat{j}(\exp_p^\gamma) = g_\#$. The transversality follows from a general theorem of Abraham by considering the map { metrics γ on M } $\times T_pM \to L(T^*M)$ given by $(\gamma, y) \mapsto \hat{j}(\exp_p^\gamma)(y)$.

If n = 3, critical points of corank 2 have codimension 3 (= codim $\Sigma_3 \cap S(R^3)$) and so are isolated. Some results on their distribution have been obtained.

A notion of stability of quasi-exponential mappings is easy to set up, using symplectic diffeomorphisms of T^*M to define equivalence. It seems probable that exponential maps are stable if and only if they are infinitesimally stable.

(25) Periodic points, measures and Axiom A R. Bowen

If $f \in \text{Diff}(M)$ satisfies Axiom A then the non-wandering set $\Omega = \Omega(f)$ decomposes into components $\{\Omega_i\}$ on each of which f is topologically transitive. Now we know ergodicity $\Rightarrow$ topological transitivity, so we ask

(1) Is it possible to define an ergodic measure (> 0 on open sets) for Ω_i?

(2) Can we say something about the distribution of periodic points on each Ω_i?

For measure μ the entropy $h_\mu(f)$ can be defined, and also the topological entropy [1] independent of μ. As in [1] we can ask (on each Ω_i):

(3) (i) Is $h_\mu(f) \leq h(f)$? (ii) $h(f) = \sup_\mu h_\mu(f)$? (iii) Does there exist unique μ with $h_\mu(f) = h(f)$?

Question (3) (i) - (iii) was answered affirmatively by W. Parry in the 0-dimensional case; the methods used can now be generalized to attack (1) - (3) in general.

Since $f|\Omega_i$ is expansive (i.e. $\exists\ \varepsilon > 0$ such that given $x, y \in \Omega_i$, $x \neq y$, $\exists\ n \in Z$ with $d(f^n x, f^n y) > \varepsilon$) we know from independent results of Shub, Walters and Bowen that $\overline{\lim}\ \frac{1}{n} \log N_n(f) \leq h(f) < \infty$, where $N_n(f)$ = number of periodic points of $f|\Omega_i$ with period n. In fact we have $\overline{\lim}\ \frac{1}{n} \log N_n(f) = h(f)$ for Axiom A.

If the stable and unstable manifolds of all points in Ω_i are dense in Ω_i, we say $f|\Omega_i$ is C-dense. In any case $\Omega_i = X_1 \cup \dots \cup X_k$, the X_j being disjoint closed subsets with $f(X_j) = X_{j+1}$, $f(X_k) = X_1$ and $f|X_j$ C-dense, so for problems of entropy etc. we assume $f|\Omega_i$ to be C-dense.

Let Ψ be a basis for the open sets of Ω_i; define $\omega_n(B) = N_n(B) / N_n(f)$ where $B \in \Psi$ and $N_n(B)$ means the number of periodic points (period n) in B. Then ω can be used to define a measure on Ω_i, and if Ψ is countable there exists an increasing sequence $\{n_k\}$ such that $\lim_{k\to\infty} \omega_{n_k}(B)$ exists for all B. This gives an invariant measure $\mu_{\{n_k\},f}$ on Ω_i, positive on open sets and zero on points. In fact the measures for all $\{n_k\}$ are equivalent! It can also be shown that f is ergodic with respect to all of them, whence by a theorem in ergodic theory they are all the same. Call them all μ_f. It is easy to deduce:

(i) U open, $\mu_f(\partial U) = 0 \Rightarrow \lim N_n(U) / N_n(f) = \mu_f(U)$,

(ii) K closed, $\mu_f(K) = 0 \Rightarrow \lim N_n(K) / N_n(f) = 0$.

THEOREM. In 0 dimensions μ_f is the unique measure maximizing entropy. In the nilmanifold case μ_f = usual Haar measure. In general $h_{\mu_f}(f) = h(f)$, and $h_\rho(f) \leq h(f)$ for any other normalised invariant measure ρ. The uniqueness of μ_f with this property has not yet been proved.

Question: Is $f|\Omega_i$ a K-automorphism with respect to μ_f when $f|\Omega_i$ is C-dense?

REFERENCES:

[1] Adler, R.L., Konheim, A.G., McAndrew, M.H. Topological entropy, Trans. AMS. 114 (1965) 309 - 319.

(26) Holomorphic vector fields on CP^2 — J. Guckenheimer

It is a classical problem to find the behaviour of solutions of $\frac{dy}{dx} = P/Q$, where P and Q are polynomials in real variables x and y. One method is to extend the field of directions given by the equation to a field on S^2 by pasting two hemispheres together. This does not go over to the complex case since the analogous identification functions cannot be made analytic. However, by writing $x = X/Z$, $y = Y/Z$ we obtain

$$R(X,Y,Z)\, dX + S(X,Y,Z)\, dY + T(X,Y,Z)\, dZ = 0$$

at each point $(X,Y,Z) \varepsilon CP^2$, and $XR + YS + ZT \equiv 0$, so we can think of the equation as a field of line elements on CP^2 by identifying tangents at $(X,Y,Z) \varepsilon CP^2$ with lines perpendicular to the vector $(X,Y,Z) \varepsilon \mathbb{C}^3$.

The same treatment does not work for the vector field $\frac{dx}{dt} = P$, $\frac{dy}{dt} = Q$. Instead, we put $u = \frac{1}{x}$, $v = \frac{y}{x}$ and find that the equations $(\frac{x}{y})^{n-1} \frac{dx}{dt} = P$, $(\frac{x}{y})^{n-1} \frac{dy}{dt} = Q$ transform to equations of the same form. They represent a section of the tangent bundle TCP^2 (or TRP^2) tensored with a line bundle $\underline{\ell}$, and a 'solution' is no longer a flow $M \times R \to R$ but a map $\phi : \underline{\ell}^{-1} \to M$.

A general holomorphic vector field on CP^2 can be expressed as $X\dot{Y} - Y\dot{X} = XB - YA$, $Y\dot{Z} - Z\dot{Y} = YC - ZB$, $Z\dot{X} - X\dot{Z} = ZA - XC$ with A,B,C homogeneous polynomials in (X,Y,Z) (see [1]). The singularities are thus the eigenvectors of the matrix $\begin{pmatrix} A \\ B \\ C \end{pmatrix}$. For an open dense set of vector fields the eigenvalues are distinct (say a_0, a_1, a_2) and the field is conjugate to one in normal form: $X\dot{Y} - Y\dot{X} = (a_1 - a_0)XY$, $Y\dot{Z} - Z\dot{Y} = (a_2 - a_1)YZ$, $Z\dot{X} - X\dot{Z} = (a_0 - a_2)ZX$. Adding a constant to all the a_i does not alter the equations, and multiplying by a scaler factor does not alter the orbits, so we can assume $a_0 + a_1 + a_2 = 0$ and the equations to be $\frac{d}{dt}(X,Y,Z) = (X, cY, -(1+c)Z)$. Thus, to investigate structural stability we need only observe the effect of perturbation of the complex parameter c.

<u>THEOREM</u>. If c is not real, the field $(X, cY, -(1+c)Z)$ is structurally stable. Hence, an open dense set of holomorphic vector fields on CP^2 are structurally stable.

<u>Corollary</u> (using above method). A generic singularity (i.e. dis-

tinct eigenvalues) of a 2-dimensional complex holomorphic vector field is locally structurally stable.

A similar theorem for CP^n probably holds.

REFERENCES:

[1] Morikawa, H. On osculating systems of differential equations of type (N,1,2), Nagoya Math. J. 31 (1968) 251-278.

(27) Small delays don't matter J. Kurzweil

An ordinary differential equation with time delay may be of the form $\dot{x} = f(x(t),x(t-1))$ ($x \in R^k$, $f : R^k \times R^k \to R^k$ smooth or Lipschitz), or more generally $\dot{x}(t) = f(x_t)$ ($x_t : [-1,0] \to R^k$ defined by $x_t(s) = x(t+s)$; $f : \mathcal{C}_{[-1,0]} \to R^k$ where $\mathcal{C}_{[-1,0]}$= continuous functions $[-1,0] \to R^k$). This can be translated into the language of dynamical systems by defining a semi-flow ψ on $\mathcal{C}_{[-1,0]}$ as follows: given the above equation and some $\phi \in \mathcal{C}_{[-1,0]}$ there is a unique 'solution' $\bar{x}(t)$ with $\bar{x}(t) = \phi(t)$ when $t \in [-1,0]$; for any $T \geq 0$ define $\psi_T(\phi)$ to be $\bar{x}_T \in \mathcal{C}_{[-1,0]}$. A subset $A \subset \mathcal{C}_{[-1,0]}$ is invariant if for all $\phi \in A$ the semi-flow ψ can be extended backwards, i.e. there exists ϕ_1 with $\psi_T(\phi_1) = \phi$ for all $T \geq 0$. We can pose the following general problem:

What are the invariant sets for given equations of the above type?

There do exist some general theorems. Consider the two equations (1) $\dot{x}(t) = g(x(t)) + h(x_t)$ and (2) $\dot{x}(t) = g(x(t)) + hp(x(t))$ (where $g:R^k \to R^k$, $p : R^k \to \mathcal{C}_{[-1,0]}$, $h : \mathcal{C}_{[-1,0]} \to R^k$).

THEOREM 1 [1]. If g is bounded and Dg is bounded and uniformly continuous, there exists $\varepsilon > 0$ such that $|h|, |Dh| \leq \varepsilon \Rightarrow$ there exists $p : R^k \to \mathcal{C}_{[-1,0]}$ with the property that any solution of (1) which is defined on the whole of R is a solution of (2), and any solution of (2) is a solution of (1). Moreover, if $\hat{g} : R^k \to \mathcal{C}_{[-1,0]}$

is defined by $\hat{g}(x) = \bar{x}|[-1,0]$ where $\bar{x}$ is the solution of $\dot{x}(t) = g(x(t))$ with $\bar{x}(0) = x$, then p and $\hat{g}$ are C^1-close.

Thus, the maximal invariant set is precisely $p(R^k)$.

The theorem can be generalized to manifolds. Let M be a Hausdorff space with a covering family of open sets $\{U_i\}_{i\varepsilon I}$ and homeomorphisms $\phi_i : U_i \to Y$ where Y is some Banach space. Call $\{(U_i,\phi_i)\}$ a <u>uniform Lipschitzian atlas of order k</u> if there are $R > 0$, $K > 0$ such that

(i) given x, there is $i \varepsilon I$ such that the ball of centre $\phi_i(x)$ and radius R is contained in $\phi_i(U_i)$,

(ii) $|| D^\ell \phi_j \phi_i^{-1} || \leq K$ $(\ell=1,2,\ldots,k)$ and $|| D^k\phi_j\phi_i^{-1}(u) - D^k\phi_j\phi_i^{-1}(v) || \leq K || u-v ||$.

Two such atlases are <u>equivalent</u> if their union is one, and equivalence classes define a <u>U.L.k-structure</u> on M which also induces a smooth structure and a metric on M.

Take $k = 1$, and $g : M \to TM$ such that the 'fibre part' $\tilde{g}_i$ is bounded by some fixed K_1 for all local products. Let $\bar{M}$ = {continuous maps $[-1,0] \to M$} (a metric space since M is a metric space).

<u>THEOREM 2</u>. There exists $\varepsilon > 0$ such that for any $h : \bar{M} \to TM$ with all $||\tilde{h}_i(x)|| < \varepsilon$ and ε a Lipschitz constant for h, there exists $p : M \to \bar{M}$ such that any solution of (1) defined on R is a solution of (2), and any solution of (2) is a solution of (1). Moreover, p and $\hat{g}$ (as in Theorem 1) are C^0-close.

<u>Applications</u>: (1) For R^k with the usual structure even the linear case $\dot{x}(t) = Ax(t) + Bx(t-1)$ cannot be treated by Theorem 1 since Ax is not bounded. By defining $U_0 = \{x \mid ||x|| < 2\}$ and $(||y|| \geq 1)$ $U_y = \{x \mid ||x-y|| < \frac{1}{2} ||y||\}$, $\phi_y(x) = (x-y) / ||y||$ we obtain a new U.L.1-structure in which a modification of Theorem 2

applies. A full analysis of the solutions can be given.

(2) $\dot{x}(t) = f(x(t), x(t-\varepsilon))$ where $f : M \times M \to TM$ is C^1 and $f(x,y) \in T_xM$ (M compact C^2). Putting $t = \varepsilon\tau$ gives $\dot{x}(t) = \varepsilon f(x(\tau), x(\tau-1))$ so if ε is small enough by Theorem 2 there exists $p : M \to \bar{M}$ with $\dot{x}(t) = \varepsilon f(p(x(\tau))(0), p(x(\tau))(-1) = f(p(x(t))(0), p(x(t))(-1))$. Now $g \equiv 0$ so $\hat{g}(x)(u) \equiv x$ and the equation is close (C^1, in fact) to $\dot{x}(t) = f(x(t), x(t))$. Thus, small delays don't matter.

REFERENCES:

[1] Kurzweil, J. Invariant manifolds for flows, Proceedings of Symposium on Differential Equations and Dynamical Systems (Puerto Rico), Academic Press, New York, 1967.

(28) Conjugacy and ergodicity

P. Walters

Let K^n denote the n-torus (usual Haar measure), and let $T : K^n \to K^n$ be an affine transformation, i.e. $T(x) = a + Ax$ where $a \in K^n$ and A is an automorphism of K^n. The following results are known:

(1) (Hoare and Parry). T is ergodic if and only if the closed subgroup of K^n generated by a and $(A-I)K^n$ is all of K^n and A has no eigenvalues which are proper roots of unity; and (Walters) this occurs if and only if T is topologically transitive. Hence A is ergodic if and only if it has no roots of unity as eigenvalues.

(2) (Hahn). T is weak mixing $\Leftrightarrow$ T is strong mixing $\Leftrightarrow$ A is ergodic.

A homeomorphism $T : X \to X$ of a compact metric space has discrete centralizer (d.c.) $\Leftrightarrow$ there exists $\delta > 0$ such that the identity I is the only homeomorphism ϕ of X with $\phi T = T\phi$ and $d(\phi, I) < \delta$. The d.c. property is clearly a topological conjugacy invariant. If for all $x,y \in X$ there exists $n \in Z$ with $d(T^n x, T^n y) > \varepsilon$ (i.e. T is expansive) then T has d.c., with constant $\delta = \varepsilon$. Also, T has d.c. $\Leftrightarrow$ there is a neighbourhood U of the diagonal Δ in $X \times X$ such that

Δ is the only $T \times T$ - invariant graph $\subset U$ of a homeomorphism of X.

THEOREM 1. If T is an invertible affine transformation on K^n, then T has d.c. $\Leftrightarrow$ T is weak mixing.

Corollary 1. A has d.c. $\Leftrightarrow$ A is weak mixing $\Leftrightarrow$ A is ergodic (by (2) (although there are ergodic automorphisms which are not expansive). (Note: Adler and Palais showed A ergodic $\Rightarrow$ every continuous mapping commuting with A is affine. In fact, A not ergodic $\Rightarrow$ there are non-affine homeomorphisms of K^n arbitrarily close to I, commuting with A.)

Proof of Cor. 1 : $\Rightarrow$ follows from the note above. To show $\Leftarrow$, suppose $\phi A = A\phi$ with ϕ near I. By the note, ϕ is affine $= B + b$, say, so $B = I$ if $\phi - I$ small enough. So $A(b) = b$. But ergodic A has only a finite number of fixed points (this follows since $(A-I)$ is onto) so if b is small enough we must have $b = 0$.

More generally, let $T = A.a$ be an invertible affine transformation of a nilmanifold $N/\Gamma = \{\Gamma x \mid x \in N\}$.

THEOREM 2. T has d.c. $\Leftrightarrow$ (i) T is ergodic and (ii) dE_e does not have $+1$ as an eigenvalue (where $E : N \to N$ defined by $x \mapsto a^{-1}.A(x).a$).

Corollary 2. A has d.c. $\Leftrightarrow$ A is ergodic and dA_e does not have $+1$ as an eigenvalue.

Define $T : X \to X$ to have unstable centralizer (u.c.) if there exists $\delta > 0$ such that I is the only continuous map $\phi : X \to X$ with $\phi T = T\phi$ and $d(\phi, I) < \delta$. The above theorems hold with u.c. replacing d.c.

For further details and more general results see [1].

REFERENCES:

[1] Walters, P. Homeomorphisms with discrete centralizers and ergodic properties, (to appear).

(29) SL(n,R) actions C. Schneider

An action of a Lie group G on a smooth manifold M is a continuous homomorphism $G \to \mathrm{Diff}(M)$. The cases G = Z or R and G compact have been studied quite successfully, although not much is known about other non-compact cases. We investigate actions of SL(n,R) (n×n real matrices of determinant +1) on manifolds of dimension $\leq 2n-3$, by combining knowledge about actions of the maximal compact subgroup SO(n) with results on R-actions (flows).

For compact groups we have [1]:-

(1) Slice theorem. Each orbit has a tubular neighbourhood on which the group action is differentiably conjugate to the induced action on the normal bundle of the orbit.

Thus, at a fixed point the action is conjugate to a linear action. This was conjectured by Palais and Smale to be true for C^∞ actions of semi-simple (non-compact) groups; we prove it to be the case for our SL(n,R)-actions.

(2) Principal orbit theorem. Let $G_x = \{g \in G \mid g(x) = x\}$ and partially order the conjugacy classes $\{(G_x)\}_{x \in M}$ by $(G_x) \leq (G_y)$ $\Leftrightarrow$ there is $g \in G$ with $gG_xg^{-1} \subset G_y$. Then there is a minimal class (H), and $\{x \in M \mid (G_x) = (H)\}$ is connected, open and dense.

Let A(G,M) denote the set of actions of G on M; let W be a manifold of dimension $\leq n-2$, and let H be as above for G = SO(n). The main step in the investigation of SL(n,R)-actions is the construction of a bijection $E \leftrightarrow A(R,W) \times Z_2$, where E is the set of SL(n,R)-actions on $SO(n)/H \times W$ which extend a certain natural SO(n)-action.

We topologize A(G,M) as the inverse limit of $\left\{A(G,M) \cap C^0(G,\mathrm{Diff}^r(M))\right\}$ where $C^0(G,\mathrm{Diff}^r(M))$ is the space of continuous maps $G \to \mathrm{Diff}^r(M)$ with the compact-open topology, and then use known generic properties of flows on W in particular cases (e.g. Peixoto's theorem [2] for dim W = 2) to deduce generic properties of SL(n,R)-actions. For example:

THEOREM 1. Generically, $\phi \in A(SL(n,R),M)$ has only a finite number of closed orbits apart from fixed points (dim $M \leq 2n-3$). The fixed point set has a finite number of components.

THEOREM 2. A dense set of intransitive SL(n,R)-actions on M^n are topological conjugacy stable.

THEOREM 3. A dense set of SL(n,R)-actions on M^{n+1} are structurally stable.

REFERENCES:

[1] Borel, A. et.al. Seminar on transformation groups, Princeton, 1950.

[2] Peixoto, M. Structural stability on two dimensional manifolds, Topology 1 (1962) 101-120.

(30) Diff(M) is simple? D. Epstein

Let M be a connected possibly non-compact C^∞ manifold, and denote by DM the group of C^1-diffeomorphisms f of M of compact support (i.e. f = identity outside some compact set in M) and isotopic to the identity by an isotopy supported in some compact set in M. Topologize DM by the fine C^1-topology.

THEOREM 1. The commutator subgroup [DM,DM] is simple. (Note [DM,DM] is always normal.)

THEOREM 2. [DM,DM] is dense in DM.

THEOREM 3. Every normal subgroup of DM contains [DM,DM].

Corollary. DM has no non-trivial closed normal subgroup.

(Theorems 1 and 3 are valid for C^r-diffeomorphisms, $1 \leq r \leq \infty$.)

THEOREM 4. Let PLM be the analogous group to DM, defined using piecewise linear homeomorphisms. If dim M = 1 then PLM is simple. The result is true for dim M = n if and only if the following conjecture is true : every $f \in$ PLM is of the form $f_1 f_2 \ldots f_k$ where each f_i is supported on a ball B^n and there consists of a 'linear shift' on an interval I and the identity on an (n-2)-sphere S^{n-2}, extended linearly to B^n = join (I, S^{n-2}).

As contrast, we observe that the group of C^∞ diffeomorphisms of R^n keeping a point fixed is not simple : $\{f \mid$ k-jet of f = identity$\}$ is a normal subgroup for each k. Also, $\{f \mid \infty$-jet of f = identity$\}$ has a strictly increasing uncountable chain of normal subgroups $(n \geq 2)$.

<u>Outline of proofs</u>. Let $U \subset M$ be an open ball, let $f \in DM$ be supported on U, and choose any $k \in DM$ with $\bar{U} \cap k^{-1}\bar{U} = \emptyset$. Take a ball $W \supset \bar{U} \cup k^{-1}\bar{U}$ and define $F \in DM$ with $F = k$ on $k^{-1}\bar{U}$, = identity outside W. Suppose $G \triangleleft DM$, $g \in G$ ($g \neq$ identity). Let Y be a ball with $\bar{Y} \cap g^{-1}\bar{Y} = \emptyset$, and let $h \in DM$ be such that $h(Y) = W$. Then $hgh^{-1} \in G$ and so $[F, hgh^{-1}] \in G$ (where $[f_1, f_2]$ means $f_2^{-1} f_1^{-1} f_2 f_1$) and $\hat{f} = [f,[F,hgh^{-1}]] \in G$. Now $\hat{f} = f$ on U and $k^{-1}f^{-1}k$ on $k^{-1}U$, and is the identity elsewhere. If $f_1 f_2$ are supported on U and k_1, k_2 are chosen with $\bar{U}, k_1^{-1}\bar{U}, k_2^{-1}\bar{U}$ disjoint it follows that $[f_1, f_2] = [\hat{f}_1, \hat{f}_2] \in G$. It is a standard result that any $f \in DM$ is a product $f_1 \ldots f_r$ with f_i, f_j supported either on disjoint balls or on a common ball; hence $[DM, DM] \subset G$, which is Theorem 3. A similar, but more elaborate, argument proves Theorem 1.

For Theorem 2, first observe that if $f = f_1 f_2$ supported on U and k_1, k_2 are as above, then the diffeomorphism

(f on U, $k_i^{-1} f_i^{-1} k_i$ on $k_i^{-1}U$ (i=1,2), identity elsewhere) is $\hat{f}$ (using k_1) followed by $\widehat{k_1^{-1} f_2 k_1}$ (using $k_1^{-1}k_2$), hence belongs to [DM,DM]. Choose $\theta \in DM$ near the identity. By considering geodesics near $x \in M$ we can write $\theta = \theta_1\theta_2$, where θ_2 takes x "half way" to $\boldsymbol{\theta}(x)$, say; similarly $\theta_1 = \theta_{11}\theta_{12}$, $\theta_2 = \theta_{21}\theta_{22}$, etc., and we have $\theta_{i_1 \cdots i_n} \to$ identity as $n \to \infty$. Suppose θ supported on a ball U contained in some coordinate neighbourhood V with origin $\notin$ U. It is sufficient to prove $\theta \in \overline{[DM,DM]}$. Let $t : V \to V$ be multiplication by $\frac{1}{2}$. Let $\phi = \theta$ on U, $= t\theta_1{}^{-1}t^{-1}$ on tU, $= t^2\theta_2{}^{-1}t^{-2}$ on t^2U, $= t^3\theta_1 t^{-3}$ on t^3U, etc.: symbolically, we write

$$\phi = \theta\ \theta_1{}^{-1}\theta_2{}^{-1}\theta_1\theta_{11}{}^{-1}\theta_{12}{}^{-1}\theta_2\theta_{21}{}^{-1}\theta_{22}{}^{-1}\theta_{11} \cdots .$$

By the above, $\phi \in \overline{[DM,DM]}$. Also, cancelling another way, $\theta^{-1}\phi \in \overline{[DM,DM]}$. Hence, $\theta \in \overline{[DM,DM]}$. (Note ϕ is C^1 but not C^2 at the origin of V).

(31) Distributed parameters control — R. Conti

Let X be a reflexive Banach space, $V \subset X$, $W : [0,T] \to$ {subsets of X}, U a Banach space, $\mathcal{L}_p(0,T\ ;U)$ the $\mathcal{L}_p$ U-valued functions on [0,T], and $\mathcal{U} \subset \mathcal{L}_p(0,T\ ;U)$ $(1 < p < \infty)$. We consider the linear control process

(1) $$\dot{x} - A(t)x = B(t)u(t)\ , \qquad t \in [0,T]$$

where $u(t) \in \mathcal{U}$ and (i) A is a function $[0,T] \to$ {linear, possibly unbounded, operators on X} which admits an evolution operator G (see [2]), (ii) $B \in \mathcal{L}_{p'}(0,T : L[U,X])$ where $1/p + 1/p' = 1$, (iii) there is a unique solution $x(t,u,v)$ of (1) with $x(0,u,v) = v$, represented by $x(t,u,v) = G(t,0)v + \int_0^t G(t,s)B(s)u(s)ds$. These conditions are satisfied in e.g. ordinary differential equations and some heat-transfer equations.

We want to find a least $t^* \in [0,T]$ $(t^* \neq 0)$ such that there exist $u^* \in \mathcal{U}$, $v^* \in V$ with $x(t^*,u^*,v^*) \in W(t^*)$.

(The finite-dimensional case with V = one point is dealt with in [3]).

For $t \in [0,T]$ we define operators $\Gamma_t : X \to X$ given by $\Gamma_t(x) = G(t,0)v$ and $\Lambda_t : \mathcal{L}_p(0,T;U) \to X$ given by $\Lambda_t(u) = \int_0^t G(t,s)B(s)u(s)ds$, and let S_t denote the Minkowski sum $S_t = -W(t) + \Gamma_t V + \Gamma_t \mathcal{U}$. Then $0 \in S_t$ is equivalent to the existence of $t \in [0,T]$, $u \in \mathcal{U}$, $v \in V$ with $x(t,u,v) \in W(t)$, i.e. the process is <u>controllable</u> using $\mathcal{U}$ and V. Now assume $\mathcal{U}$, V and W(t) are bounded, and define $h(t,x') = \sup_{x \in S_t} \langle x,x' \rangle$ (x' $\in$ conjugate X' of X). Using a separation theorem for convex sets we can show [1]:

<u>THEOREM A</u> (controllability). Let (i)-(iii) above be satisfied, let V,W(t) (each $t \in [0,T]$) be convex, closed and bounded, let $\mathcal{U}$ be convex, bounded and with $\Lambda_t \mathcal{U}$ closed for each $t \in [0,T]$. Then the process is controllable at time t if and only if $h(t,x') \geq 0$ for all $x' \in X'$.

<u>THEOREM B</u> (existence). Let the assumptions of Theorem A be satisfied, and let W(t) be a continuous (in the space of subsets of X, with Hausdorff metric) function of t. Then, if the process is controllable using $\mathcal{U}$,V it is also controllable in a minimum time.

The following theorem, which takes the rôle of Pontryagin's maximal principle, is proved in [2]:

<u>THEOREM C</u> (maximum principle). Let the assumptions of Theorem B be satisfied, and suppose $V \cap W(0) = \emptyset$. Then, if t^*,u^*,v^* is a solution of the minimum time problem, there exists $x_0' \in X'$, $\underline{x_0' \neq 0}$, with (i) $\langle x(t^*,u^*,v^*),-x_0' \rangle = \sup_{W(t^*)} \langle w,-x_0' \rangle$, (ii) $\langle v^*,G'(t,0)x_0' \rangle = \sup_V \langle v,G'(t^*,0)x_0' \rangle$ and (iii) $\int_0^{t^*} \langle B(s)u^*(s),G'(t^*,s)x_0' \rangle ds = \sup_{\mathcal{U}} \int_0^{t^*} \langle B(s)u(s),G'(t^*,s)x_0' \rangle ds$ where G' is the conjugate of G, <u>provided</u> (a) $\dim X < \infty$ <u>or</u> (b) $\operatorname{int} W(t^*) \neq \emptyset$ <u>or</u> (c) $\operatorname{int} V \neq \emptyset$

and Γ_t is onto (all $t \in [0,t^*]$) or (d) int $\mathcal{U} \neq \emptyset$ and Λ_t is onto (all $t \in [0,t^*]$).

An example due to Egorov shows (a)-(d) cannot all be dispensed with.

REFERENCES:

[1] Conti, R. On some aspects of linear control theory, Mathematical Theory of Control (ed. Balakrishnan and Neustadt), Academic Press, New York, 1967.

[2] Conti, R. Time-optimal solution of a linear evolution equation in Banach spaces, J. Optimization Theory and Applications. 2 no. 5 (1968) 277-284.

[3] Lee, E.B. and Markus, L. Foundations of Optimal Control Theory, J. Wiley and Sons, New York 1967.

(32) Flows outside a compact invariant set — T. Saito

Let $\Pi : X \times R \to X$ be a flow on a locally compact metric space X (the local compactness is necessary, the metric property may not be). Define $C^\pm(x), L^\pm(x)$ to be the $\pm$ orbits and limit sets of x (see p.23), and let F be a compact invariant set. We study the orbits near F. For the case when F is an isolated minimal set see [2].

Let $U \supset F$ be a relatively compact neighbourhood, and define $G_U = \{x \mid x \in \bar{U}-F,\ C^+(x) \not\subset \bar{U},\ C^-(x) \not\subset \bar{U}\}$, $N_U^+ = \{x \mid x \in \bar{U}-F,\ C^+(x) \subset \bar{U}\}$, $N_U^- = \{x \mid x \in \bar{U}-F,\ C^-(x) \subset \bar{U}\}$, $N_U = N_U^+ \cap N_U^-$. (These generalize Bendixson's definitions [1] with $X = R^2$, F an isolated critical point, and the flow C^1). Note G_U is open, and N_U is closed in $\bar{U}-F$ and invariant. We assume F to be not open, and isolated from minimal sets, so we can suppose $\bar{U}-F$ contains no minimal sets.

THEOREM 1 (Ura-Kimura [4], [5]). If $\partial U \neq \emptyset$ then neither $N_U \cup F$ nor

$G_U \cup F$ contains any neighbourhood of F.

THEOREM 2. If $N_U = \emptyset$ then $x \in N_U^{\pm} \Rightarrow L^{\pm}(x) \subset F$.

THEOREM 3. $N_U^{\pm} - N_U$ contains no ± Poisson stable points (i.e. x with $x \in L^{\pm}(x)$).

If $\overline{G_U} \cap F \neq \emptyset$ for some U (hence for all small enough U) call F a saddle set (Seibert, Ura). For $x \in X$ define the positive prolongational limit set $J^+(x) = \{y \mid \exists \{x_n\} \subset X, \{t_n\} \subset R$ with $x_n \to x, t_n \to \infty, \Pi(x_n, t_n) \to y\}$ (Auslander, Bhatia and Seibert), and $J^-(x)$ similarly using $t_n \to -\infty$.

THEOREM 4. F is a saddle set if and only if there exists $x \in X - F$ with either $L^+(x) \cap F \neq \emptyset$, $J^+(x) \cap (X-F) \neq \emptyset$ or $L^-(x) \cap F \neq \emptyset$, $J^-(x) \cap (X-F) \neq \emptyset$.

Corollary. F not a saddle set, $L^+(x) \cap F \neq \emptyset \Rightarrow L^+(x) \subset F$.

THEOREM 5. F not a saddle set $\Rightarrow$ there exists a neighbourhood $V \supset F$ such that $V - F$ is disjoint from all Poisson stable orbits.

THEOREM 6. F is asymptotically stable whenever it is stable.

THEOREM 7. F is ±-vely stable $\Leftrightarrow N_U^{\mp}$ is empty, for some U.

Proofs of Theorems 4,6 and 7 can be found in [3]. Results have also been obtained on 'attractor' theory (see [3]), and on the structure of $\bar{U} - F$ when F is not a saddle set (unpublished).

REFERENCES:

[1] Bendixson, I. Sur les courbes définies par des équations différentielles, Acta. Math. 24 (1901) 1-88.

[2] Saito, T. Isolated minimal sets, Funkcial. Ekvac. 11 (1968) 155-167.

[3] Saito, T. On a compact invariant set isolated from minimal sets, (to appear in Funkcial. Ekvac. 12 (1969)).

[4] Ura, T. and Kimura, I. Sur le courant extérieur à une région invariante; théorème de Bendixson, Comm.Math.Univ.Sancti Pauli 8 (1960) 23-39.

[5] Ura, T. On the flow outside a closed invariant set, Contr. Diff. Equns. III (1964) 249-294.

(33) Non-linear Volterra equations J. Nohel

We consider systems of Volterra equations of the form

$$\text{(I)}\qquad x(t) = f(t) - \int_0^t a(t,s).G(s,x(s))ds$$

where $0 \leq s \leq t < \infty$, $f : R^+ \to R^n$, $G : R^+ \times R^n \to R^n$ and $\underline{a}$ is an $n \times n$ matrix. Alternatively, we use the form

$$\text{(I')}\qquad x'(t) = \zeta(t,x(t)) - \int_0^t A(t,s)G(s,x(s))ds \qquad (x(0)=x_0).$$

Questions can be asked on the following: (i) existence and uniqueness of solutions (locally easy, by usual methods, but globally non-trivial), (ii) boundedness of solutions, (iii) (asymptotic) stability of solutions, (iv) existence of asymptotically periodic or almost periodic solutions. See references ([1]-[7]).

Equation (I) can be written $x(t) = y(t) - \int_0^t r(t,s)h(s,x(s))ds$ where $h = G - x$, $y(t) = f(t) - \int_0^t r(t,s)f(s)ds$ is the solution of the linear problem $y(t) = f(t) - \int_0^t a(t,s)y(s)ds$, and r satisfies the "resolvent equation" $r(t,s) = a(t,s) - \int_s^t a(t,u)r(u,s)du$.

THEOREM 1. Assume (1) $r(t,s) \in \mathcal{L}_1(\text{loc})$ $(0 \leq s \leq t)$, (2) $h(t,x)$ measurable in (t,x) $(0 \leq t < \infty, \ ||x|| < \infty)$, $h(t,0) \equiv 0$, (3) given $\varepsilon > 0$ there exists δ such that

$$||x||, ||y|| \leq \delta \Rightarrow ||h(t,x) - h(t,y)|| \leq \varepsilon ||x-y||,$$ (4) $\sup_{t\geq 0} \int_0^t ||r(t,s)||ds \leq B$, some $B > 0$. Then given $\varepsilon > 0$ there exists $\delta > 0$ such that $||f||_\infty \leq \delta \Rightarrow$ there is a unique solution $x(t)$ of (I) in $\mathcal{L}_\infty(0,\infty)$ with $||x||_\infty \leq \varepsilon$ (i.e. we have global existence, uniqueness and stability). If

$$\lim_{h\to 0}\left(\int_t^{t+h}||r(t+h,s)||ds + \int_0^t ||r(t+h,s) - r(t,s)||ds\right) = 0$$

the solution $x(t)$ is bounded and continuous. If $||f||_{BC} \to 0$ and if for any $T > 0$ $\lim_{t\to\infty}\int_0^T ||r(t,s)||ds = 0$, then $\lim_{t\to\infty} ||x(t)||_{BC} = 0$. (Here $||\ ||$ is any consistent vector or matrix norm, $||\ ||_\infty$ is the $\mathcal{L}_\infty$ norm and $||\ ||_{BC}$ is the uniform norm).

If f is real and periodic (period ω), $r(t+\omega,s+\omega) = r(t,s)$, and $\lim_{t\to\infty}\int_{-\infty}^0 ||r(t,s)||ds = 0$, then y is asymptotically periodic (i.e. $\exists$ a function p such that $\lim_{t\to\infty} ||y(t)-p(t)|| = 0$, where $p(t+\omega) = p(t)$).

THEOREM 2. If in addition the solution of the linear system $y(t)$ is asymptotically periodic and $g(t,x)$ is periodic in t there is a bounded continuous solution $x(t)$ on $[0,\infty)$ which is asymptotically periodic. (The behaviour of x can be described in terms of y in more general cases).

We now turn to (I').

THEOREM 3. Consider $x'(t) = F(t) - \int_0^t a(t-s)g(x(s))ds$ (here $n = 1$), $x(0) = x_0$. Assume

(1) $a(t)$ continuous $(0\le t<\infty)$, $(-1)^k a^{(k)}(t)\ge 0$ $(0<t<\infty)(k=0,1,2)$,

(2) $xg(x) \ge 0$, $\int_0^x g(\xi)d\xi \to \infty$ as $|x| \to \infty$,

(3) $|g(x)| \le K(1 + \int_0^x g(\xi)d\xi)$, (4) $F(t)$ continuous, $\in \mathcal{L}_1[0,\infty)$.

Then a solution exists and is bounded. If also $a(t) \not\equiv a(0)$, $a'''(t) \le 0$, $xg(x) > 0$ $(x \neq 0)$, $g'(x)$ is continuous and $|F'(t)|$ is bounded at least for t sufficiently large then every solution and its derivative $\to 0$ as $t \to \infty$.

Theorem 3 has been generalized by Levin (non-convolution case) [1] and Vinokurov (systems) [7].

REFERENCES:

[1] Levin, J.J. A non-linear Volterra equation, not of convolution type, J.Diff.Equations 4 (1968) 176-186.

[2] Levin, J.J. and J. Nohel. Perturbation of a non-linear Volterra equation, Mich.Math.J. 12 (1965) 431-447

[3] Levin, J.J. and J. Nohel. On a non-linear delay equation, J.Math. Analysis and Applications 8 (1964) 31-44.

[4] Miller, R.K., J. Nohel and J.S.W. Wong. Perturbations of Volterra integral equations, J.Math.Analysis and Applications 25 (1969) 676-691.

[5] Miller, R.K. and G.R. Sell. Existence, uniqueness and continuity of solutions of integral equations, Ann. Mat.Pura.Appl. 80 (1968) 135-152.

[6] Nohel, J. Some problems in non-linear Volterra integral equations, Bull.A.M.S. 68 (1962) 323-329.

[7] Vinokurov, V.R. Asymptotic behaviour of a class of integro-differential Volterra equations, Diff.Urav. 3 (1967) 1732-1744.

(34) Ergodic Hamiltonian theory

L. Markus

Let M^{2n} be a symplectic manifold (with closed non-singular 2-form Ω), and let $H : M \to R$ be a Hamiltonian defining a Hamiltonian flow on M in the standard way. The following are traditional conjectures for generic Hamiltonian systems: (I) the flow is ergodic on each energy manifold ($H^{-1}(h)$, h constant) (II) there is no integral (i.e. $F : M \to R$ invariant under the flow) independent of H. In the 1930's, Oxtoby and Ulam showed measure-preserving topological flows to be generically ergodic. We show that (I) is false for Hamiltonian flows, and (II) is true (n=2). Let $\mathfrak{H}$ be the set of Hamiltonians on M (defined up to constants), given the C^∞ topology. Define $H \in \mathfrak{H}$ to be ergodic when the flow on $\{x \in M | H(x) = h\}$ is ergodic on at least one component, for a dense set of $h \in R$.

THEOREM 1. Let M^{2n} be a compact symplectic manifold. There is an open dense set $\mathcal{G} \subset \mathcal{H}$ consisting of Hamiltonians which are not ergodic.

Define $F : M \to R$ to be independent of H if dF and dH are linearly independent almost everywhere (using the measure μ given by Ω^n) on M.

THEOREM 2. If M is as above with n = 2, there is a residual set $\mathcal{R} \subset \mathcal{H}$ such that each $H \in \mathcal{R}$ has no C^∞ integral independent of H.

Remark. The theorems still hold if we consider non-compact $M^{2n} = T^*M^n$ with $\mathcal{H} = \mathcal{H}_V$ consisting of H of the form $T + V\pi$ where $\pi : T^*M \to M$ and T is a fixed Riemannian metric. These results are joint work with K. Meyer.

Sketch of proof of Theorem 2. By a theorem of Robinson there is a residual set $\mathcal{R} \subset \mathcal{H}$ such that for $H \in \mathcal{R}$ (1) each critical point has eigenvalues linearly independent over Z (2) except for countably many periodic orbits the nontrivial characteristic multipliers are $\neq 1$ (3) H has a unique absolute minimum on M and there has a generic elliptic critical point (eigenvalues $i\omega_1, \ldots, i\omega_n$, $-i\omega_1, \ldots -i\omega_n$ (independent $\omega_j > 0$) and a quadratic non-degeneracy condition). Take $H \in \mathcal{R}$, and let $S^3_h = H^{-1}(h)$ (h small). There is a family of invariant tori T^2_h in S^3_h, filling a set of μ_H-measure > 0 in S^3_h (Arnold, Moser). Suppose F is an integral independent of H: then on some $S^3_{h_0}$ consider the restricted function F_h and there $dF_h(p) \neq 0$ for some $p \in T^2_{h_0}$ and hence $\neq 0$ in a neighbourhood of p. It can then be shown that dF_h is nonzero on the whole of $T^2_{h_0}$. An implicit function theorem gives the existence of a continuous family of invariant tori $T^2_{f,h_0} = (F \times H)^{-1}(f, h_0)$ filling a $T^2 \times [0,1]$. The rotation number changes continuously and does change (using the quadratic non-degeneracy condition) so there exists a rational rotation number on some $T^2_{f_0,h_0}$. Then $T^2_{f_0,h_0}$ contains a periodic orbit, and is filled with periodic orbits (since μ_H is preserved), contradicting (2).

(35) Subharmonic solutions to Duffing's equation M. Urabe

For physical systems described by a weakly nonlinear Duffing's equation one can prove, using the method of averaging, that, if the damping is absent or very small, subharmonic oscillations only of order 1/3 can appear only when the frequency of the oscillating external force is close to a particular value. When the last condition is satisfied, six subharmonic oscillations of order 1/3 appear, but three of them are not stable and the remaining three are stable or have neutral stability according as the damping is present or absent.

For systems described by a strongly nonlinear Duffing's equation, by the use of the numerical results obtained by Galerkin's procedure and the speaker's existence theorem [1,2], one can see that six subharmonic oscillations of order 1/3 can really appear, and the same conclusions as in the weakly nonlinear case can be obtained concerning the stability of the subharmonic oscillations. A remarkable character of these subharmonic oscillations is that the first two terms strongly dominate the remaining terms in their Fourier-series expansions. This implies that even for the strongly nonlinear case one can determine the qualitative character of subharmonic oscillations by investigating approximate solutions involving only two leading harmonic terms.

REFERENCES:

[1] Urabe, M. Galerkin's procedure for nonlinear periodic systems, Arch.Rat.Mech.Anal. 20 (1965) 120-152.

[2] Urabe, M. and A. Reiter. Numerical computation of nonlinear forced oscillations by Galerkin's procedure, J.Math.Analysis and Applications 14 (1966) 107-140.

(36) Similarity of automorphisms of the torus R. Adler

Let ϕ, ϕ' be measure-preserving transformations of normalized measure spaces $(X,\mathcal{B},m)$, $(X',\mathcal{B}',m')$ respectively, and let $\theta : X \to X'$ be invertible and satisfy $\theta\phi = \phi'\theta$ and $m' = m\theta^{-1}$: then ϕ is said to be metrically similar to ϕ'. Write $\phi \sim \phi'$. If h (or h_m) denotes entropy, it is easy to show $\phi \sim \phi' \Rightarrow h_m(\phi) = h_{m'}(\phi')$. The converse is in general false (take $\phi : x \mapsto x + \gamma \bmod 1$ on unit interval, ϕ' = identity), but we seek conditions under which it may hold. We say ϕ is a Kolmogorov transformation if there exists $\mathcal{A} \subseteq \mathcal{B}$ with (i) $\mathcal{A} \subseteq \phi\mathcal{A}$, (ii) $\bigvee_{-\infty}^{\infty} \phi^n(\mathcal{A}) = \mathcal{B}$, (iii) $\bigcap \phi^n(\mathcal{A}) = \{\emptyset, X\}$. Examples include geodesic flows on surfaces of (constant) negative curvature, symbolic shifts, and continuous ergodic automorphisms on compact separable abelian groups.

Conjecture. If ϕ, ϕ' are Kolmogorov then $h_m(\phi) = h_{m'}(\phi') \Rightarrow \phi \sim \phi'$.

MAIN THEOREM. (Adler, Weiss [1]). If ϕ, ϕ' are continuous ergodic automorphisms on the torus R^2/Z^2 then $h(\phi) = h(\phi') \Rightarrow \phi \sim \phi'$.

The proof goes by showing ϕ and ϕ' are both similar to 'symbolic shifts' which are similar to each other. Let $\mathcal{A} = \{1,\ldots,N\}$, let $T = (t_{ij})$ be an N×N matrix of 0's and 1's, and define $\Xi(T) = \{\xi \mid \xi = (\ldots\xi_{-1}\xi_0\xi_1\ldots),\ \xi_n \in \mathcal{A}$ and $t_{\xi_n\xi_{n+1}} = 1$ for all $n\}$. Define $\sigma : \Xi(T) \to \Xi(T)$ to be a shift if $(\sigma\xi)_n = \xi_{n+1}$. There are many σ-invariant measures that can be defined on $\Xi(T)$; in particular, Parry [2] defines a measure μ with the following properties: (1) $h_\mu(\sigma) = \log \lambda_T$, where λ_T is the eigenvalue of T of greatest modulus, (2) for any measure m, $h_m(\sigma) \leq h_\mu(\sigma)$ with equality $\Leftrightarrow m = \mu$. Now, for any $\phi : X \to X$ we can define

$\tau : X \to E(T)$ for suitable T by choosing a partition $\gamma = C_1 \cup C_2 \cup \ldots \cup C_N$ of X and letting $\tau(x) = (\ldots \xi_{-1}(x), \xi_0(x), \xi_1(x) \ldots)$ where $\phi^n(x) \in C_{\xi_n(x)}$, and $t_{ij} = 1$ if and only if $\phi C_i \cap C_j \neq \emptyset$.

We prove the following:

THEOREM 2. Suppose (i) γ is a generator (i.e. $\bigvee_{-\infty}^{\infty} T^n\gamma = \mathcal{B}$), (ii) T is irreducible (i.e. given i,j there exists n with $(T^n)_{ij} > 0$), (iii) $h_m(\phi) = \log \lambda_T$. Then $\phi \sim \sigma$ (via τ).

In the case of the main Theorem, the transformations ϕ, ϕ' are represented by 2 × 2 integer matrices of determinant ± 1. The geometrical nature of these transformations shows that the hypothesis of Theorem 2 is satisfied. This leads to a representation of the automorphisms of the torus by symbolic shifts. It then becomes a combinatorial problem to show that shifts arising from automorphisms of equal entropy are metrically similar.

REFERENCES:

[1] Adler, R. and B. Weiss. Entropy, a complete metric invariant for automorphisms of the torus, Proc. Nat.Acad.Sci.U.S.A. 57 (1967) 1573-1576.

[2] Parry, W. Intrinsic Markov chains, Trans.A.M.S. 112 (1964) 55-66.

(37) Differential equations with periodic coefficients B.Schmitt

We consider $\frac{dx}{dt} = f(t,x)$ where $x \in$ euclidean n-space E_n, $t \in R$, and $f(t+1,x) = f(t,x)$ (but not excluding $f(t+\alpha,x) = f(t,x)$, $\alpha<1$). Assume f such that there exists a unique solution $x(t,t_0,x^0)$ with $x = x^0$ when $t = t_0$, continuous in $t,(t_0,x^0)$ and defined for all $t \in R$.

Define $T : E_n \times S^1 \to E_n \times S^1$ by $T(x^0,t_0) = (x(t_0+1,t_0,x^0),t_0)$. The return curve of x_0 is $p_1T\,(x^0 \times S^1)$, where $p_1 : E_n \times S^1 \to E_n$ is projection on the first factor.

Take $n = 2$. Define $x^0 \in E_2$ to be <u>exceptional</u> if x^0 lies on a periodic solution of period 1; let $N(f) = \{x^0 \mid x^0 \text{ exceptional}\}$. If $x^0 \notin N(f)$ define the <u>index</u> $i(f,x^0) \in Z$ (called 'rotation number' by H. Seifert) to be the index of the return curve of x^0 with respect to x^0. (Note that in the autonomous case $i(f,x^0) = 0$ since the return curve is a point). In each component of $E_2 - N(f)$ the index is constant: hence if $x^0, y^0 \notin N(f)$ and $i(f,x^0) \neq i(f,y^0)$ we know x^0 and y^0 are separated by solutions of period 1. In fact, there is the <u>Seifert formula</u>:

$$i(f,y^0) = i(f,x^0) + \sum_\nu \gamma_\nu S(f_\nu,q)$$

where q is a path x^0y^0, f_ν is a periodic solution (isolated, period 1), γ_ν is its fixed-point index, and S is the intersection number of f_ν and q.

<u>Applications</u>. (1) $x'' + x^3 = \sin t$. Write $x_1' = x_2$, $x_2' = -x_1^3 + \sin t$. Computor calculations give (suppressing f) $i(0,0) = -1$, $i(2,0) = 0$: hence between (2,0) and (0,0) there pass periodic solutions of period 1. In fact, there is only one such, and the Seifert formula gives $0 = -1 - \gamma$, i.e. $\gamma = -1$, showing the periodic solution is semi-stable.

(2) $x'' + x + x^3 = \sin t$. Here $i(0,0) = -1$, but the return curve passes too close to (2,0) for $i(2,0)$ to have been calculated.

Suppose $N(f)$ bounded: then $i(f,x^0)$ is the same for all large x^0 and is called $i(f)$. Let $y(t) = D_t x(t)$ where $t \in S^1$, $D_0 = D_1 =$ identity, D_t is a diffeomorphism and $D_t(x)$ is differentiable in t (if we consider only <u>flows</u>, the differentiability is unnecessary). We obtain an <u>equivalent</u> system $\frac{dy}{dt} = g(t,y)$, and $i(g)$ is defined.

<u>Proposition</u>. $i(g) = i(f) + k(1-\gamma)$ where γ = fixed point index of

$N(f)$, k = linking number of $D_t(x^0)$, $D_t(y^0)$ (with $E_2 \times S^1$ embedded as a 'solid torus' in E_3).

Suppose $f(t,a^0) = 0$ for all t (i.e. a^0 is a constant trajectory). Suppose that there is a neighbourhood U of a^0 such that the index is defined for f and its linearization ℓf for $x \in U$.

THEOREM. $i(f,x) = i(\ell f,x)$, $x \in U$.

Consider $f(t,x) = A(t)x$ (A = 2×2 matrix). Suppose that $x = 0$ is the only solution of period 1; $\gamma = \pm 1$. A theorem of Floquet states that the system is equivalent to one of the form $\frac{dy}{dt} = By$ (autonomous); this has zero index, so $i(f) = 0$ ($\gamma = 1$) or $i(f) = 2k$ ($\gamma = -1$). Reality problems may arise if $\gamma = 1$; a direct argument is then needed [1]. We ask: is an arbitrary system equivalent to one with $i(f) = 0$? The answer is yes when we can linearize. Also, a theorem of Seifert states that if the maximum diameter of the trajectories in $E_2 \times [0,1]$ is finite then $i(f) = 0$.

REFERENCES:

[1] Schmitt, B.V. Index associé à un système différentiel linéaire, périodique, du second ordre, l'Enseignment Mathématique 13 (1967) 313-323.

(38) An algebraic approach to dynamical systems R. Ellis

A dynamical system means a group T acting on a compact Hausdorff space X, the action written $(x,t) \mapsto xt$ with $(xt)s = x(ts)$, $xe = x$. Let $\mathcal{P}$ denote point-transitive systems (X,T) (i.e. for which X contains a dense orbit), let $\mathcal{M}$ denote systems for which X is a minimal set (i.e. every orbit is dense). Our approach is to study the whole classes $\mathcal{P}$, $\mathcal{M}$ rather than systems in isolation.

In $\mathcal{P}$ and $\mathcal{M}$ there exist "universal objects": there is a unique (up to isomorphism) $(P,T) \in \mathcal{P}$ ($(M,T) \in \mathcal{M}$) such that any

$(X,T) \in \mathcal{P}$ $((Y,T) \in \mathcal{M})$ is the image of (P,T) $((M,T))$ under an equivariant map. In fact, P is βT = (Stone-Čech) β-compactification of T; βT has a natural semi-group structure, and T can be thought of as a dense subgroup of βT. T acts on βT point-transitively by right multiplication. A set $M \subset \beta T$ is minimal $\Leftrightarrow$ M is a minimal right ideal, and all such are isomorphic: this is the universal minimal set $M \in \mathcal{M}$.

Let $C(X)$ = complex-valued continuous functions on X. Any $\phi: (\beta T,T) \to (X,T)$ induces $\phi^* : C(X) \to C(\beta T) = \mathcal{C}$, say. Given $f \in \mathcal{C}$ define functions fx, xf by $fx(y) = f(xy)$, $xf(y) = f(yx)$ (fx is continuous, but xf may not be unless $x \in T$). These give right and left actions of T on $\mathcal{C}$. We now look at <u>T-subalgebras</u> of $\mathcal{C}$, i.e. subalgebras $\mathcal{A}$ closed in the uniform norm, and with $f \in \mathcal{A}$, $t \in T \Rightarrow \bar{f} \in \mathcal{A}$, $tf \in \mathcal{A}$. Note $\phi^*(C(X))$ is a T-subalgebra. Thus we can study dynamical systems by studying T-subalgebras of $\mathcal{C}$. Given $\mathcal{A}$, let $|\mathcal{A}| = \mathrm{Hom}_T(\mathcal{A},\mathcal{C})$, define $R_s : \mathcal{C} \to \mathcal{C}$ by $f \mapsto fs$ $(s \in T)$, and define an action of T on $|\mathcal{A}|$ by $\Psi s = R_s \circ \Psi$ $(\Psi \in |\mathcal{A}|$, $s \in T)$. Then in fact $(|\mathcal{A}|,T)$ is the system with $\phi^*(C|\mathcal{A}|)) = \mathcal{A}$.

Any $\Psi : \mathcal{A} \to \mathcal{C}$ is a restriction of some $\hat{\Psi} : \mathcal{C} \to \mathcal{C}$, so given (X,T) represented by $\mathrm{Hom}_T(\mathcal{A},\mathcal{C})$ and (Y,T) represented by $\mathrm{Hom}_T(\mathcal{B},\mathcal{C})$ and $\Psi \in \mathrm{Hom}_T(\mathcal{A},\mathcal{C})$ we can study $\hat{\Psi}|\mathcal{B}$. Thus we can consider "how X acts on Y".

Let M be a minimal set in βT, with J = {idempotents in M}: then (1) $vx = x$ $(v \in J,\ x \in M)$, (2) $M = \sum_{v \in J} Mv$, (3) Mv is a group with identity v. Now $\mathcal{A}$ represents a minimal set $\Leftrightarrow$ $\mathcal{A} \subset \mathfrak{A}(v)$ for some $v \in J$ $(\mathfrak{A}(v) = \{f | fv = v\})$, and in fact we can take v fixed, $= u$, say. Let $G = Mu$. Given $\mathcal{A} \subset \mathfrak{A}(u)$ define $\mathcal{G}(\mathcal{A}) = \{\alpha | \alpha \in G,\ f\alpha = f\ (f \in \mathcal{A})\}$. It turns out that $\mathcal{G}(\mathcal{A})$ is a subgroup of G. Given $H \subset G$ there is in general no $\mathcal{A}$ with $\mathcal{G}(\mathcal{A}) = H$, but for a certain topology τ on G any <u>closed</u> H does correspond to some $\mathfrak{A}$. We observe that $\mathcal{G}(\mathfrak{A}) = \mathcal{G}(\mathfrak{B}) \not\Rightarrow \mathfrak{A} = \mathfrak{B}$.

Define $\mathfrak{B}$ to be a distal extension of $\mathfrak{A}$ (write $\mathfrak{A} \leq \mathfrak{B}$) when (i) $\mathfrak{A} \subset \mathfrak{B}$ and (ii) if a homomorphism p = pw on $\mathfrak{A}$ then p = pw on $\mathfrak{B}$, for all $w \in J$. (Called 'distal' because $(|\mathfrak{A}|,T)$ is distal $\Leftrightarrow \mathbb{C} \leq \mathfrak{A}$). Write $A = \mathcal{G}(\mathfrak{A})$, $B = \mathcal{G}(\mathfrak{B})$. In general $B \not\lhd A$; if $B \lhd A$ there is a group H of homeomorphisms of $|\mathfrak{A}|$ such that $(|\mathfrak{A}|,T) = (|\mathfrak{B}|/H,T)$.

If $\mathfrak{A} \leq \mathfrak{B}$ let $[\mathfrak{A},\mathfrak{B}] = \{\mathfrak{F} \mid \mathfrak{A} \subset \mathfrak{F} \subset \mathfrak{B}\}$, $[A,B] = \{H \mid B \subset H \subset A,\ H \text{ closed}\}$. There is a bijection $\phi : [\mathfrak{A},\mathfrak{B}] \to [A,B]$ given by $\phi(\mathfrak{F}) = \mathcal{G}(\mathfrak{F})$, with inverse $H \mapsto \mathfrak{A}(H) \cap \mathfrak{B}$ $(\mathfrak{A}(H) = \{f \mid f\alpha = f,\ \alpha \in H\})$.

Given $\mathfrak{F} \subset \mathcal{C}$, there is a unique maximal distal extension $\mathfrak{F}^*$ (i.e. $\mathfrak{F} \leq \mathfrak{A} \Leftrightarrow \mathfrak{F} \subset \mathfrak{A}$ and $\mathfrak{A} \subset \mathfrak{F}^*$). Let $F = \mathcal{G}(\mathfrak{F})$, and define $f : F \to \mathbb{C}$ to be almost periodic over $\mathfrak{F}$ if $f \in \mathfrak{F}^*$ and the maps $\alpha \mapsto f(\alpha p): F \to \mathbb{C}$ are continuous with respect to a topology τ^* (similar to above) $(p \in M)$: when $\mathfrak{F} = \mathbb{C}$ we get the usual definition. Then $\mathfrak{A}$ is an almost periodic extension of $\mathfrak{F}$ (write $\mathfrak{F} \leq_p \mathfrak{A}$) if $f \in \mathfrak{A} \Rightarrow f$ is almost periodic over $\mathfrak{F}$. Let $F^\# = \mathcal{G}(\mathfrak{F}^\#)$ where $\mathfrak{F}^\# = \{f \mid f \text{ a.p. over } \mathfrak{F}\}$; then in fact $F^\#$ is the intersection of the closed neighbourhoods of e in F.

Suppose $\mathfrak{F} \leq \mathfrak{A}$, $\mathfrak{F} \neq \mathfrak{A}$ and $|\mathfrak{A}|$ metrizable. Furstenberg's theory [1] gives the existence of $\mathfrak{B}$ with $\mathfrak{F} \neq \mathfrak{B}$ and $\mathfrak{F} \leq_p \mathfrak{B} \subset \mathfrak{A}$. In the case $\mathfrak{F} = \mathbb{C}$ the proof depends on finding a neighbourhood V of e in $G = \mathcal{G}(\mathbb{C})$ with $\bar{V} \neq G$, which is done by constructing a non-constant semi-continuous function ϕ on G, e.g. $\phi(\alpha) = \inf_{t \in T} d(ut,\alpha t)$ (where $\mathfrak{A} \subset \mathfrak{A}(u)$). Note ϕ is not constant since $(|\mathfrak{A}|,T)$ is distal; it would suffice for $(|\mathfrak{A}|,T)$ to be point-distal i.e. $\exists\ x_0$ such that $\inf d(x_0t,yt) = 0 \Rightarrow x_0 = y$.

For $\mathfrak{F} \leq_p \mathfrak{F}^\#$ as above define a T-$\mathfrak{F}$ submodule of $\mathfrak{F}^\#$ to be a subspace $\mathfrak{M} \subset \mathfrak{F}^\#$ such that $g \in \mathfrak{M} \Rightarrow fg \in \mathfrak{M}$, $tg \in \mathfrak{M}$

($f \varepsilon \mathfrak{F}, t \varepsilon T$). The map $\kappa : |\mathfrak{F}^{\#}| \to |\mathfrak{F}|$ induced by $\mathfrak{F} \subset \mathfrak{F}^{\#}$ makes $|\mathfrak{F}^{\#}|$ into a fibre space with fibre $H = F/F^{\#}$. If we assume $\mathcal{M}_a = \{f$ restricted to $\kappa^{-1}(a)$, $f \varepsilon \mathcal{M}, a \varepsilon |\mathfrak{F}|\}$ to be finite-dimensional, then dim $\mathcal{M}_a$ turns out to be independent of a. A <u>T-vector bundle</u> ξ over $|\mathfrak{F}|$ is a vector bundle $\pi : E(\xi) \to B(\xi)$ with $B(\xi) = |\mathfrak{F}|$ and π a homomorphism commuting with T-actions on E and B, also with an inner product on $E(\xi)$ giving a metric under which $\pi^{-1}(a) \to \pi^{-1}(at)$ is an isometry. Thus $\kappa : |\mathfrak{F}^{\#}| \to |\mathfrak{F}|$ is a T-vector bundle. All T-vector bundles arise in this way; in fact, there are bijections {T- $\mathfrak{F}$ submodules} $\leftrightarrow$ {T-vector bundles} $\leftrightarrow$ {unitary representations of H}.

REFERENCES:

[1] Furstenberg, H. The structure of distal flows, <u>Amer. J.Math</u>. 85 (1963) 477-515.

(39) <u>Volterra equations and semi-flows</u> <u>J. Nohel</u>

In (<u>33</u>) (p. 58) a Volterra equation of the form

$$x(t) = f(t) + \int_0^t g(t,s,x(s))ds \tag{1}$$

was discussed. Here we interpret (1) in the language of semi-flows, and use properties of semi-flows to give some information about solutions of (1).

Let $\alpha : X \to R^+$ be a lower semi-continuous function on a metric space X, let $I(p) = [0,\alpha(p))$ ($p \varepsilon X$), $S = \{(t,p) \mid t \varepsilon I(p), p \varepsilon X\}$. A local <u>semi-flow</u> on X is a continuous function $\pi : S \to X$ such that (i) $\pi(0,p) = p$, and $\pi(s,\pi(t,p)) = \pi(s+t,p)$ ($s, t \geq 0$) whenever $s+t \in I(p)$, and (ii) $I(p)$ is maximal, i.e. $\overline{\{\pi(t,p) \mid t \varepsilon I(p)\}}$ is not compact in X. Define $X(\infty) = p \varepsilon X \mid \alpha(p) = +\infty\}$, $\gamma(p) = \{\pi(t,p) \mid t \varepsilon I(p)\}$, and call a subset $A \subset X$ <u>invariant</u> if $A \subset X(\infty)$ and if the orbit $\gamma(p) \subset A$ for all $p \varepsilon A$.

THEOREM 1 (Invariance principle). If $\overline{\gamma(p)}$ is compact then $p \in X(\infty)$ and the ω-limit set of p ($= \{q \mid \pi(t_m,p) \to q$ as $m \to \infty$ for some sequence $\{t_m\} \to \infty\}$) is nonempty, invariant, compact and connected.

Let C denote the set of continuous functions $R^+ \to R^n$, topologized by seminorms $||\phi||_J = \sup_{t\in J}|\phi(t)|$ (J compact interval $\subset R^+$); let $\mathcal{G}_C$ denote a certain collection (suitably topologized) of functions g such that (1) has a unique continuous solution on $0 \leq t < \alpha(f,g)$ for each $f \in C$ and $[0,\alpha(f,g))$ is maximal. Define $S = \{(t;f,g) \mid t \in [0,\alpha(f,g)),\ f \in G,\ g \in \mathcal{G}_C\}$, and define $T : S \to C$ by $T(t;f,g)(\theta) = \int_0^t g(t+\theta;s,x(s;f,g))ds$ (where x is the solution of (1)). Define $f_t \in C$ by $f_t(\theta) = f(t+\theta)$ and $g_t \in \mathcal{G}_C$ by $g_t(u,v,x) = g(t+u,t+v,x)$. Finally, let $X = C \times \mathcal{G}_C$ and define $\pi : S \to X$ by

$$\pi(t;f,g) = (f_t + T\,(t;f,g),g_t).$$

THEOREM 2. π is a local semi-flow on X.

In order to apply Theorem 1 it suffices to show:

THEOREM 3. Suppose $\{g_t \mid t \geq 0\}$ has compact closure in $\mathcal{G}_C$ and (1) has unique solution x(t) bounded and uniformly continuous on $[0,\infty)$. Then $\{\pi(t;f,g) \mid t \geq 0\}$ has compact closure in $C \times \mathcal{G}_C$.

Given $\{t_n\} \to \infty$, there is a subsequence $\{t_k\}$ and functions $X^* \in C$, $g^* \in \mathcal{G}_C$ such that $\{x_{t_k}\} \to x^*$ (where $x_{t_k}(\theta) = x(t_k + \theta)$), $g_{t_k} \to g^*$; then $f_{t_k} \to f^*$ where $f^*(\theta) = x^*(\theta) - \int_0^\theta g^*(\theta,z,x^*(z))dz$. Note x^* is a solution which can be continued to the whole of $(-\infty,\infty)$ for the limiting equation

$$x^*(t) = f^*(t) + \int_0^t g^*(t,s,x^*(s))ds.$$

An important special case is when $g(t,s,x) = a(t-s)h(s,x)$, where $a \in \mathcal{L}_1(R^+)$ and h is independent of s (or periodic in s) and Lipschitz in x. For example, using the invariance principle we can prove:

THEOREM 4 (Levin, [1]). Suppose f continuous, $f(t) \to$ constant $(=f(\infty))$ as $t \to \infty$, $a \in \mathcal{L}_1(R^+)$, $a(t) \geq 0$, $a'(t) \leq 0$, $a'(t) \not\equiv 0$ on any $[0,\varepsilon]$ $(\varepsilon>0)$, $h(x) = e^x-1$. Let x_0 be the unique solution of $x = f(\infty) + h(x) \int_0^\infty a(s)ds$. Then a unique solution $x(t)$ of (1) exists on $[0,\infty)$, is bounded, and $\lim_{t\to\infty} x(t) = x_0$.

Interesting problems arise when f and h are periodic in t. It can be shown that the limiting equation has a periodic solution, but it is not known whether it is unique.

REFERENCES:

[1] Levin, J.J. The qualitative behaviour of a non-linear Volterra equation, Proc. A.M.S. 16 (1965) 711-718.

[2] Miller, R.K. Nonlinear Volterra equations and topological dynamics, (to appear in: Advances in Differential and Integral Equations, SIAM (1969) (Proceedings of Conference held in Madison, Wis., August 1968)).

[3] Miller, R.K. Asymptotic behaviour of solutions of nonlinear Volterra equations, Bull. A.M.S. 72 (1966) 153-157.

[4] Miller, R.K. and G.R. Sell. A note on Volterra integral equations and topological dynamics, Bull.A.M.S. 74 (1968) 904-908.

[5] Miller, R.K. and G.R. Sell. Existence, uniqueness and continuity of solutions of integral equations, Ann.Mat.Pura Appl. 80 (1968) 135-152.

[6] Nohel, J.A. Some problems in nonlinear Volterra integral equations, Bull.A.M.S. 68 (1962) 323-329.

[7] Sell, G.R. Nonautonomous differential equations and topological dynamics I, Trans. A.M.S. 127 (1967) 241-262.

(40) Homomorphisms of minimal sets

J. Auslander

We consider transformation groups (X,T) where T is a fixed group acting on a compact Hausdorff space X. A minimal set in

X is a non-empty closed T-invariant set containing no proper subset with these properties. If the closure of the orbit of $x \in X$ is minimal then x is <u>almost periodic</u>.

There is a unique (up to isomorphism) <u>universal</u> minimal transformation group (M,T), i.e. if (X,T) is minimal (X a minimal set) then there is a homomorphism (equivariant continuous map) $(M,T) \to (X,T)$. A useful property of (M,T) is: if $m, m' \in M$ such that (m,m') is an almost periodic point of $(M \times M, T)$ then there is an automorphism α of (M,T) with $\alpha(m) = m'$.

Let $G = \text{Aut}\,(M,T)$; let (M,T) be minimal with $\gamma:(M,T) \to (X,T)$. We study $G(X,\gamma) = \{\alpha \in G \mid \gamma\alpha = \gamma\}$, a subgroup of G.

<u>Lemma</u>. If $\gamma, \gamma' : M \to X$ there exists $\beta \in G$ with $\gamma' = \gamma\beta$.

Thus $G(X,\gamma') = \beta^{-1} G(X,\gamma)\beta$; so every minimal set determines a conjugacy class of subgroups of G.

Given $\pi : X \to Y$ (omitting T) we have $G(X,\gamma) \subset G(Y,\pi\gamma)$. If $G(X,\gamma) \neq G(Y,\pi\gamma)$ say π is <u>proper</u>.

<u>THEOREM 1</u>. Let (X,T),(Y,T) be minimal, with $\pi : X \to Y$ a homomorphism. The following are equivalent:

(i) π is proper

(ii) there exist $x, x' \in X$, $x \neq x'$, $\pi(x) = \pi(x')$, such that (x,x') is an almost periodic point of $(X \times X, T)$,

(iii) there exist $x, x' \in X$, $x \neq x'$, $\pi(x) = \pi(x')$, such that x, x' are <u>not</u> proximal. (x, x' are <u>proximal</u> if there is a net $\{t_n\} \subset T$, $z \in X$, such that $xt_n \to z$ and $x't_n \to z$).

<u>Corollary</u>. If (X,T),(Y,T) distal minimal, then π not proper $\Rightarrow$ π an isomorphism.

<u>THEOREM 2</u>. Let $\gamma : M \to X$, $\gamma' : M \to X'$ be homomorphisms. Then

(i) $G(X,\gamma) \subset G(X',\gamma') \Leftrightarrow$ there is minimal (Y,T) with homomorphisms $\pi : Y \to X$, $\pi' : Y \to X'$ where π improper,

(ii) $G(X,\gamma) = G(X',\gamma') \Leftrightarrow$ there is minimal (Y,T) with improper $\pi : Y \to X$, $\pi' : Y \to X'$.

(iii) Suppose (X,T), (X',T) are distal. Then $G(X,\gamma) \subset G(X',\gamma')$ $(G(X,\gamma) = G(X',\gamma'))$ $\Leftrightarrow$ there is a homomorphism (isomorphism) $\pi : X \to X'$ with $\gamma' = \gamma\pi$.

Define $\pi : X \to Y$ to be of <u>distal type</u> if

$$x, x' \in X,\ x \neq x',\ \pi(x) = \pi(x') \Rightarrow x, x' \text{ not proximal.}$$

<u>THEOREM 3</u>. The following are equivalent:

(i) π is of distal type,

(ii) $\pi(x) = \pi(x') \Rightarrow (x,x')$ is an almost periodic point of $(X \times X, T)$,

(iii) if $y \in Y$ then $\pi^{-1}(y)$ is an almost periodic set,

(iv) if $\gamma : M \to X$ and $x \in X$ then

$$\pi^{-1}\pi(x) = \bigcup \{\gamma\alpha\gamma^{-1}(x) \mid \alpha \in G(Y, \pi\gamma)\}.$$

<u>THEOREM 4</u>. If $\pi : X \to Y$ is of distal type then π is open, and all the sets $\pi^{-1}(y)$ have the same cardinality.

(41) <u>Boundedness of solutions of 2nd order equations</u> — <u>Dame Mary Cartwright</u>

We consider the equation

$$\ddot{x} + kf(x)\dot{x} + g(x) = 0 \qquad (k > 0),$$

and seek conditions under which the solution $x(t;x_0,y_0)$ with $x_0 = x(0;x_0,y_0)$, $y_0 = \dot{x}(0;x_0,y_0)$ satisfies

$$|x(t;x_0,y_0)| < B,\ |\dot{x}(t;x_0,y_0)| < B(k+1)\ (t > t_0(k,x_0,y_0,B))$$

where B depends on f,g but not on x_0,y_0,k. This boundedness of solutions then implies the existence of periodic solutions.

The equation is equivalent to the system $\dot{x} = y-kF(x)$, $\dot{y} = -g(x)$ where $F(x) = \int_0^x f(\xi)d\xi$. This can be studied using a form of energy equation

$$V(x,y) = G(x)+\tfrac{1}{2}y^2 = V(x_0,y_0)-k\int_0^t F(x(t))g(x(t))dt$$

obtained by multiplying the first equation by g(x) and the second by y, adding, and integrating.

The system in the form

$$\frac{dy}{dx} = -\frac{g(x)}{y-kF(x)}$$

is also useful for the strip $|x| \le 1$ (if $\mathcal{B}'$ below is used) to obtain a bound for y in terms of (x_0,y_0), and also for the strip $|x| \le B_2$ in which $|F(x)| \le B_1$.

We assume that f and g are continuous and satisfy conditions sufficient to ensure uniqueness and list some possible additional hypotheses:

$\mathcal{A}^0$: $xg(x) > 0, \quad x \neq 0$

$\mathcal{A}^1$: $xg(x) > 0, \quad |x| > 1$

$\mathcal{B}^0$: $F(x) \ge 0, \; F(-x) \le 0, \quad x \ge 0$

$\mathcal{B}^1$: $F(x) \ge 0, \; F(-x) \le 0, \quad x \ge 1$

$\mathcal{C}^+$: $G(x) \to \infty$ as $x \to \infty$

$\mathcal{C}^-$: $G(-x) \to \infty$ as $x \to \infty$

$\mathcal{D}^+$: $F(x) \to \infty$ as $x \to \infty$

$\mathcal{D}^-$: $F(-x) \to -\infty$ as $x \to \infty$

Since f,g are continuous there is a constant B_1 such that $|g| < B_1, \quad |F| \le B_1, \quad |x| \le 1$.

<u>THEOREM</u>. If $\mathcal{A}^0$, $\mathcal{B}^1$, $\mathcal{C}^+$, $\mathcal{D}^-$ hold, there exist constants $K(1+\frac{1}{k})$, $K(1+k)$ such that $-K(1+\frac{1}{k}) < x(t;x_0,y_0) < K(k+1)$, $|y| < B(k+1)$ $(t > t_0(B,k,K,x_0,y_0))$.

Similar results may be obtained with other combinations of $\mathcal{C}^+, \mathcal{C}^-, \mathcal{D}^+, \mathcal{D}^-$ and corresponding constants $K(k+1)$, $K(1+1/k)$ according as a $\mathcal{D}$ condition or a $\mathcal{C}$ condition is used, provided that <u>either</u> $\mathcal{D}^+$ or $\mathcal{D}^-$ is used and if one or other is <u>not</u> used, then $\mathscr{C}^+$ or $\mathscr{C}^-$ is used as the case may be. If $|x| <$ $\min\{K(1+1/k), K(1+k)\}$, this is equivalent to $|x| < B$. The fact that damping on one side is sufficient is due to Burton.

Similar results also hold for the equation

$$\ddot{x} + kf(x,\dot{x})\dot{x} + g(x) = 0,$$

although in this case the proof is more difficult.

(42) Möbius transformations in stability theory R.A. Smith

We consider the system

(1) $a_0 D^n x + a_1 D^{n-1} + \ldots + a_n x = 0$

where $x(t) \in \mathbb{C}^m$, $t \in R$, $D = \frac{d}{dt}$ and the complex matrices a_i are continuous functions of $t, x, Dx, \ldots, D^{n-1}x$. Also, $\det a_0 \neq 0$. A sufficient condition for the Lyapunov stability of the solution $x \equiv 0$ is the existence of an autonomous quadratic Lyapunov function (AQLF), i.e. a positive definite hermitian form $V(x, Dx, \ldots, D^{n-1}x)$ with constant coefficients such that $DV < 0$ for all solutions $x(t) \not\equiv 0$. Given a Möbius transformation $u(\xi) = (\alpha\xi + \beta)/(\gamma\xi + \delta)$, $(\alpha\delta \neq \beta\gamma)$, we define from the characteristic polynomial

$$f(\xi) = a_0\xi^n + a_1\xi^{n-1} + \ldots + a_n$$

of (1) a new polynomial $g(\xi) = (\alpha\delta - \beta\gamma)^{-n/2}(\alpha - \gamma\xi)^n f(u^{-1}(\xi))$, which is the characteristic polynomial of a new system.

(2) $b_0 D^n x + b_1 D^{n-1} x + \ldots + b_n x = 0.$

If all the roots of $\det f(\xi) = 0$ lie in $\Pi = \{\xi \in \mathbb{C} \mid \mathrm{Re}\,\xi < 0\}$ then the roots of $\det g(\xi) = 0$ lie in Π if $u(\Pi) \subset \Pi$. Thus when the a_i are constant, if $u(\Pi) \subset \Pi$ the asymptotic stability of (1) implies that of (2). When the a_i are not constant this argument fails, but the following result is true:

THEOREM 1. If $u(\Pi) \subset \Pi$ and (1) has an AQLF then (2) has an AQLF.

Replacing $D^i x$ formally by $x(t + i)(0 \leq i \leq n)$ gives a system of difference equations

(3) $a_0 x(t + n) + a_1 (t + n - 1) + \ldots + a_n x(t) = 0.$

We can define an AQLF (say W) analogous to V above, replacing the condition $DV < 0$ by $w(t + 1) < w(t)$ where

$$w(t) = W(x(t), x(t + 1), \ldots, x(t + n - 1)).$$

Replacing $D^i(x)$ by $x(t + i)$ in (2) gives

(4) $b_0 (t + n) + b_1 x(t + n - 1) + \ldots + b_n x(t) = 0.$

Now $\Delta = \{\xi \in \mathbb{C} \mid |\xi| < 1\}$ plays the role of Π. Regardless of whether the a_i are constant we have:

<u>THEOREM 2</u>. If $u(\Delta) \subset \Delta$ and (3) has an AQLF then (4) has an AQLF.

<u>THEOREM 3</u>. If $u(\Pi) = \Delta$, then (4) has an AQLF if and only if (1) has an AQLF.

A possible application is to the real scalar equation

(2*) $b_0 D^2 x + b_1 Dx + b_2 x = 0.$

Starzhinskii [1] showed that (2*) has an AQLF if $\lambda \leq b_1/b_0 \leq \Lambda$, $\mu \leq b_2/b_0 \leq M$ where λ, Λ, μ, M are positive constants satisfying certain inequalities. Applying Theorem 1 to the corresponding equation $a_0 D^2 x + a_1 Dx + a_2 x = 0$ we obtain new sufficient conditions for (2*) to have an AQLF.

Further details of the above work are to appear.

<u>REFERENCES</u>:

[1] Starzhinskii, V.M. Sufficient conditions for stability of a mechanical system with one degree of freedom, <u>Prikl. Mat. Meh</u>. 16 (1952) 369-374.

(43) <u>Some maximum principles for Itô equations</u> I. Vrkoč

We consider diffusion processes which are governed by Itô stochastic equations

(1) $dx = a(t, x)dt + B(t, x)\, dw(t)$

where $a(t, x)$ is a vector function $a(t, x) \in R^n$, $B(t, x)$ is an $n \times n$ matrix function and $w(t)$ is an n-dimensional Wiener process. Let D be a region in R^n and T be a given positive number; let Q be the 'cylindrical' region $(0, T) \times D \subset R^{n+1}$. Let $P(B,a,f,Q) = P\{\exists\ t | t \in (0,T),\ x_f(t) \notin D\}$, where P denotes probability, $x_f(t)$ is the solution of (1) with initial density f in D (such that $\int_D f(x)\,dx = 1$). A condition will be given which guarantees that the maximum of P(...) occurs in the case of the "greatest" matrix function $B(t, x)$. Before the exact formulation is given some basic notation and definitions are needed.

Definition 1. Let a vector function $a(t, x)$, a matrix function $B(t, x)$ and a region Q be given. Say a,B,Q satisfy condition (I) if:

i) The region D is bounded and the boundary $\dot{D}$ can be locally expressed by means of functions which have Hölder continuous second derivatives;

ii) The matrix function $C(t, x) = B(t, x)B^*(t, x)$ (where $B^*(t, x)$ is the transpose matrix) is positive definite in $\bar{Q}$, and $a(t, x)$, $C(t, x)$ are Hölder continuous in $\bar{Q}$;

iii) The Itô equation (1) has a unique solution for every initial density $f(x)$ in D;

iv) The parabolic equation

$$\frac{\partial u}{\partial t} = \frac{1}{2}\sum_{i,j} C_{ij}(T-t, x)\frac{\partial^2 u}{\partial x_i \partial x_j} + \sum_i a_i(T-t, x)\frac{\partial u}{\partial x_i} \tag{2}$$

has a unique bounded solution satisfying

$$u(0, x) = 0 \text{ for } x \in D,\ u(t, x) = 1 \text{ for } t > 0,\ x \in \dot{D}; \tag{3}$$

v) The bounded solution $u(t, x)$ satisfies $P(B,a,f,Q) = \int_D f(x)u(T,x)\,dx$ for every density $f(x)$ in D.

There exists a well-known condition which ensures iii): $B(t, x)$ is Lipschitz continuous in x and continuous in t. Sufficient conditions for iv) and v) can also be given.

In the following definition the main problem will be formulated precisely.

Definition 2. A matrix function $B(t, x)$ is called maximal with respect to a vector function $a(t, x)$ and a region Q if

(a) $C(t, x) = B(t, x)B^*(t, x)$ is a diagonal matrix in Q,

(b) $a(t, x)$, $B(t, x)$, Q satisfy condition (I),

(c) $P(B,a,f,Q) = \max P(B',a,f,Q)$ for all densities f in D,

where the maximum is taken over the set of matrix functions $B'(t, x)$ fulfilling conditions (a), (b), and $C'_{ii}(t, x) \leqq C_{ii}(t, x)$.

Definition 3. A matrix function $B(t, x)$ is called strongly maximal with respect to a vector function $a(t, x)$ and a region Q if condition (b) is satisfied and if

(c*) $P(B,a,f,Q) = \max P(B',a,f,Q)$ for all densities f in D, where the maximum is taken over the set of matrix functions $B'(t, x)$ satisfying condition (b), and $C(t, x) - C'(t, x)$ is a positive semi-definite matrix for every $[t, x] \in Q$.

THEOREM 1. Let $a(t, x)$, $B(t, x)$ and Q be given such that a,B,Q satisfy condition (I) and $C(t, x) = B(t, x)B^*(t, x)$ is diagonal in Q. The matrix function $B(t, x)$ is maximal with respect to $a(t, x)$ and Q if and only if the bounded solution of (2) fulfilling (3) satisfies $\frac{\partial^2 u}{\partial x_i^2} \geqq 0$, $i = 1, \dots n$, in Q.

THEOREM 2. Let $a(t, x)$, $B(t, x)$ and Q be given such that condition (I) is satisfied. The matrix function $B(t, x)$ is strongly maximal with respect to $a(t, x)$ and to Q if and only if the bounded solution of (2) satisfying (3) is a convex function with respect to x in Q.

For the one-dimensional case a stronger result is valid which cannot be generalized directly to $n > 1$.

THEOREM 3. Let a region Q be defined by $Q = (0, T) \times (x_1, x_2)$ where x_1, x_2 are numbers $x_1 < x_2$. Let a function $B(t, x)$ be defined on $\bar{Q}$ so that $B(t, x)$ is continuous in t, Lipschitz continuous in x, $B(t, x) \neq 0$ on $\bar{Q}$, and $B^2(t, x)$ and $\partial^2 B^2(t, x)/\partial x^2$ are Hölder continuous in t, x. Let $\alpha(t)$, $\beta(t)$ be Hölder continuous functions on $[0, T]$. Put $a(t, x) = \alpha(t) + \beta(t)x$. If $a(t, x_2) \leqq 0$, $a(t, x_1) \geqq 0$ then the function $B(t, x)$ is (strongly) maximal with respect to the function $a(t, x)$ and the region Q.

Under some additional conditions these results can be modified also for the case of noncylindrical regions Q. The following examples show that Theorem 3 has no direct generalization for $n > 1$.

Let $n = 2$ and let D be the square $D = (0, 1)^2$. It can be shown that the unit matrix is maximal with respect to $a \equiv 0$ and Q, but it is not strongly maximal. If the region D is an equilateral triangle then the unit matrix is not maximal with respect to $a \equiv 0$ and Q. In the case when D is a circle the unit matrix is not strongly maximal with respect to $a \equiv 0$ and Q, i.e. the formula (c*) is not valid, but the matrix function $B'(t, x)$ which gives the maximum in this formula (where B is the unit matrix) can be found.

(44) A periodic wave propagation model for pattern formation in embryos — B. Goodwin

The structure of any differentiated tissue results from a well-defined sequence of events in which the spatial and temporal organization of the developing tissue mass are intimately related. It is as though every cell has access to, and can read, a clock and a map (Wolpert's positional information). The model developed in the present paper is one in which the map arises from wave-like propagation of activity from localised clocks or pacemakers. Individual cells are supposed temporally organized in the sense that biochemical events essential for the control of development recur periodically. This temporal organization of an individual cell is converted by functional coupling between cells into a spatial ordering of the temporal organization. More explicitly, a periodic event is postulated which propagates outward from a pacemaker region, synchronizing the tissue and providing a time base for development. Intercellular signalling, entrainment of all cells in the tissue by the fastest cells in the pacemaker region, and a refractory period to guarantee unidirectional propagation are the essential features of the propagation; they permit the derivation of a wave equation and a set of boundary conditions. An underlying gradient of frequency of the event establishes the position of the pacemaker region and the sense of propagation. A second event which propagates more slowly than the first provides positional information in the form of a one-dimensional sequence of surfaces of constant phase difference between the two events. A third event is used to regulate the pattern of phase difference and thus establish size independent structures. The longest trajectory orthogonal to the surfaces of constant phase difference beginning at the pacemaker region and terminating at the regulating region defines a developmental axis of definite polarity. The model is readily extended to more than one axis, i.e. multi-dimensional positional information. It has a high informational capacity and is readily applied to the discussion of particular developmental phenomena. To illustrate its utility, we discuss development and regeneration in Hydra, positional organization in the early amphibian embryo, and the retinal-tectal projection of the amphibian visual system. Specific experiments to test for the existence of the postulated periodic events and their consequences are suggested. Some preliminary experimental results on Hydra tending to confirm the model are reported. Possible detailed realizations of the model in terms of biochemical control circuits within the cell are conjectured and discussed to show that the formal features of the model can be realized by well-recognized biochemical processes.

(45) Intrinsically ergodic systems B. Weiss

Let T be a homeomorphism of a compact space X, and let M be the set of normalized ($m(X) = 1$) T-invariant regular measures on X. If M reduces to a single element then (X, T) is said to be uniquely ergodic, a concept introduced by Krylof and Bogolyubof. Denote the entropy of T with respect to $m \in M$ by $h_m(T)$ and set $\bar{h}(T) = \sup_{m \in M} h_m(T)$.

If $\bar{h}(T) < +\infty$ and there is a unique $\bar{m} \in M$ such that $\bar{h}(T) = h_{\bar{m}}(T)$ we shall say that (X, T) is an intrinsically ergodic system (i.e.s.). The usefulness of this concept becomes clear when one realizes how the isomorphism problem simplifies for i.e.s.. Namely if (X, T) and (X', T') are i.e.s. and there exists a 1 - 1 measurable $\theta : X \to X'$ with $\theta T = T'\theta$ then the processes $(X,T,\bar{m})$ and $(X',T',\bar{m}')$ are isomorphic, that is $\theta\bar{m} = \bar{m}'$. In [1] the notion of an i.e.s. was used implicitly in this way.

We discuss now some old and new examples of i.e.s..

Subshifts of finite type are i.e.s. as was shown by Parry [4] in a different formulation. A generalization of this class has been suggested by Furstenberg as follows:

Let G be a finite semigroup with generators $a_1 \dots a_n$ and a two-sided zero ($0.g = g.0 = 0$ for all $g \in G$). A sequence (x_i), $x_i \in \{1, \dots, n\}$, is G-admissible if for all $i \leq j$

$$a_{x_i} a_{x_{i+1}} \cdots a_{x_j} \neq 0.$$

X_G is the set of all bilateral G-admissible sequences. The shift restricted to X_G is i.e. if G satisfies certain hypotheses that correspond to the aperiodicity of the transition matrix in [4].

The entropies that have occurred thus far, i.e. the numbers $\bar{h}(T)$, have been rather restricted. The same holds for the numbers that occur as entropy of automorphisms of compact groups (see [1] for the torus and [2] for the general case) which are i.e.s.. To get another class of examples where $\bar{h}(T)$ can be anything in $(0, \infty)$ proceed as follows:

Let $\{e_i^{(1)}\}$, $\{e_i^{(2)}\}$, ..., $\{e_i^{(a)}\}$ be a sequence $(i = 1,2,\ldots)$ of zeros and ones. A bilateral sequence of symbols $\{0,1,2,\ldots,a\}$ will be deemed <u>admissible</u> if

(i) the symbols $(1,2,\ldots,a)$ are isolated, that is each occurrence of $j \in (1,2,\ldots,a)$ is preceded and followed by a zero,

(ii) if the length of a block of zeros that follows $1 \leq j \leq a$ is x then $e_x^{(j)} = 1$.

The shift restricted to the closure of the set of admissible sequences is i.e.* and $\bar{h}$ is log c where c is the unique positive root of $\sum_{1}^{\infty} \left(\sum_{j=1}^{a} e_i^{(j)} \right) t^{-i} - t = 0$, and hence by an appropriate choice of the $e^{(j)}$, $\bar{h}$ may assume any value.

Dinaburg [3] has made use of these sets (for a = 1) in calculating the topological entropy of C^{∞} diffeomorphisms restricted to a part of the non-wandering set.

We end with a few <u>conjectures</u>:

---

* NOTE: A condition of <u>aperiodicity</u> must be added to ensure this fact.

1. Let X be a closed shift-invariant subset of n-ary sequence space, such that the shift is a K-automorphism on X with respect to some measure on X. Then the shift restricted to X is i.e.s..

2. In case there is a measure m such that $h_m(T) = h_{top}(T)$ (that is the topological entropy of T) then m is unique. (Assume here topological transitivity).

REFERENCES:

[1] Adler, R. and B. Weiss — Entropy, a complete metric invariant for automorhisms of the torus, Proc. Nat. Acad. Sci. 57 (1967) 1573-76.

[2] Berg, K.R. — Convolutions of invariant measures and maximal entropy,(to appear).

[3] Dinaburg, E.I. — An example of the calculation of topological entropy, Uspehi Mat. Nauk 23 (142) (1968)249-50.

[4] Parry, W. — Intrinsic markov chains, Trans. A.M.S. 112 (1964) 55-66.

(46) The group of diffeomorphisms, and motion of fluids — D. Ebin

Consider the Euler equations

$$\frac{\partial u_i}{\partial t} = \sum_j u_j \frac{\partial u_i}{\partial x^j} = \frac{\partial p}{\partial x^i} \qquad \left(\sum_i \frac{\partial u_i}{\partial x^i} = 0\right)$$

representing the motion of a non-viscous incompressible fluid, where $u(x_1, x_2, x_3; t) \in$ some domain $D \subset R^3$ and there is a given initial condition $u(x_1, x_2, x_3; 0)$. We have a differential equation on the space $C^\infty(TD)$ of C^∞ vector fields on D, but since $C^\infty(TD)$ is only a Fréchet space there is no existence theorem for solutions. Solutions were obtained in special cases by Lichtenstein in the 1920's.

Our approach is to return to the physical situation as follows. The motion of the fluid corresponds to a path in Diff(D). More generally we consider an n-dimensional Riemannian manifold M, possibly with boundary, and let $\mathcal{D}$ = Diff(M). Then the equations become: $$\frac{\partial u}{\partial t} + \nabla_u u = \text{grad } p$$

$$\text{div}(u) = 0$$

where u is a time dependent vector field on M, and "∇" is the affine connection corresponding to the Riemannian structure. The group has the structure of a differentiable manifold modelled on a Fréchet space. Let μ be a smooth volume on M, and $\mathcal{D}_\mu$ = {diffeomorphisms preserving μ} : then $\mathcal{D}_\mu$ is a subgroup of $\mathcal{D}$ and has the structure of a closed sub-manifold. This is the configuration space for the motion of perfect fluids on M. Arnol'd [1] has shown that the solutions of the Euler equations correspond to geodesics on $\mathcal{D}_\mu$ relative to a right invariant metric on $\mathcal{D}_\mu$ induced by a Riemannian structure on M. We are able to prove the existence of such a smooth geodesic flow, and hence deduce

<u>THEOREM</u>: The Euler equations have unique smooth solutions for short time-intervals, varying smoothly with the initial conditions.

We suspect but cannot yet prove that the solutions do exist for all time.

In order to prove the theorem we need to enlarge $\mathcal{D}$ to the set $\mathcal{D}^s$ of all maps $\eta : M \to M$ such that η and η^{-1} each have square integrable distribution derivatives up to order s. If $s > \frac{n}{2} + 1$ then $\mathcal{D}^s$ is a Hilbert manifold and a topological group in some sense analogous to a Lie group. The method can be modified to give solutions of the Navier Stokes equations (viscous fluids), and one finds that the solutions approach the solutions of Euler's equations as the viscosity constant approaches zero.

For full details of the above see [2,3].

<u>REFERENCES</u>:

[1] Arnol'd, V. Sur la géometrie différentielle des groupes de Lie de dimension infinie et ses applications a l'hydrodynamique des fluides parfaits, <u>Ann</u>.

Inst. Fourier, Grenoble 16 (1966) 319-361.

[2] Ebin, D.G. and J.E. Marsden — Groups of diffeomorphisms and the motion of an incompressible fluid, (to appear in Ann. Math.).

[3] Ebin, D.G. and J.E. Marsden — Groups of diffeomorphisms and the solution of the classical Euler equation for a perfect fluid, (to appear in Bull. A.M.S.).

(47) Positional information and the spatial pattern of cellular differentiation

L. Wolpert

The problem of pattern is considered in terms of how genetic information can be translated in a reliable manner to give specific and different spatial patterns of cellular differentiation. Pattern formation thus differs from molecular differentiation which is mainly concerned with the control of synthesis of specific macromolecules within cells rather than the spatial arrangement of the cells. It is suggested that there may be a universal mechanism whereby the translation of genetic information into spatial patterns of differentiation is achieved. The basis of this is a mechanism whereby the cells in a developing system may have their position specified with respect to one or more points in the system. This specification of position is positional information. Cells which have their positional information specified with respect to the same set of points constitute a field. Positional information largely determines with respect to the cells' genome and developmental history the nature of its molecular differentiation. The specification of positional information in general precedes and is independent of molecular differentiation. The concept of positional information implies a co-ordinate system and polarity is defined as the direction in which positional information is specified or measured. Rules for the specification of positional information and polarity are discussed. Pattern regulation, which is the ability of the system to form the pattern even when parts are removed, or added, and to show size invariance as in the French Flag problem, is largely dependent on the ability of the cells to change their positional information and interpret this change. These concepts are applied in some detail to early sea urchin development, hydroid regeneration, pattern formation in the insect epidermis, and the development of the chick limb. It is concluded that these concepts provide a unifying framework within which a wide variety of patterns formed from fields may be discussed, and give new meaning to classical concepts such as induction, dominance and field. The concepts direct attention towards finding mechanisms whereby position and polarity are specified, and the nature of reference points and boundaries. More specifically, it is suggested that the mechanism is required to specify the position of about 50 cells in a line, relatively reliably, in about 10 hours. The size of embryonic fields is, surprisingly, usually less than 50 cells in any direction.

(48) Bifurcations — K. Meyer

We consider one-parameter families of area-preserving diffeomorphisms of R^2. Examples of these include (i) Poincaré maps on cross-sections of periodic orbits in constant energy levels of Hamiltonian systems, (ii) period maps of the $(x,\dot{x})$ plane to itself given by equations such as Duffing's equation $\ddot{x} + x + x^3 = \varepsilon \cos \omega t$ (period $\frac{2\pi}{\omega}$, parameter ε).

Let $\{\phi_\varepsilon\}$ be such a family, where $\varepsilon \in (-\delta, \delta) \subset R$, $\phi_\varepsilon(0) = 0$ and $\det(d\phi_\varepsilon(o)) = 1$. The eigenvalues of $d\phi_\varepsilon(0)$ ("multipliers") are either both real (λ, λ^{-1}) or complex $(\lambda, \bar{\lambda} = \lambda^{-1})$. A classical result is the following:

Lemma. If for $\varepsilon = 0$ the multipliers are not +1 there is a one-parameter family of fixed-points of $\{\phi_\varepsilon\}$.

We study the case when λ is a kth root of unity ($k \geq 5$: $k = 3,4$ are special cases). Taking polar coordinates (r, θ) we write $\phi_\varepsilon(r,\theta) = (R,\Theta)$, and obtain expressions for R and Θ in terms of r, θ by 'Birkhoff's normalization'. These are analysed in two cases, called 'computable' and 'generic'. In the computable case an argument due to Birkhoff shows the existence of at least 2k fixed points of ϕ_ε^k (although possibly a continuum of them) in a neighbourhood of 0; in the generic case we obtain k fixed points which are hyperbolic (multipliers real) and k are elliptic (multipliers complex). The results in the computable case can be used to find some new periodic solutions in the 3-body problem.

The above analysis can be applied directly to Duffing's equation, written as a Hamiltonian system with $H(x,y) = \frac{1}{2}(x^2 + y^2) + x^4/4$. When $\varepsilon = 0$ the origin is a $\frac{2\pi}{\omega}$-periodic solution, and when ε small $\neq 0$ there exist at least two periodic solutions, i.e. subharmonics.

The above represents joint work with J. Palmore.

(49) <u>Théorie de Fuchs sur une variété analytique complexe</u> <u>R.Gérard</u>

Une équation différentielle linéaire d'ordre n est dite de Fuchs à l'origine de $\mathbb{C}$, si elle s'écrit dans un disque D centré en ce point sous la forme:

$$(1) \qquad x^n \frac{d^n y}{dx^n} + x^{n-1} a_1(x) \frac{d^{n-1} y}{dx^{n-1}} + \ldots + a_n(x) y = 0$$

où les $a_i(x)$ sont holomorphes dans D.

Toute solution f de (1) est holomorphe sur le revêtment universel R de $\hat{D} = D - \{0\}$ et d'ordre fini à l'origine, c'est à dire qu'il existe un nombre réel $\lambda(f)$ tel que

$$(2) \qquad \lim_{\substack{x\to 0\\ x\in U}} \frac{f(Q)}{Q^{\lambda(f)}} < \infty \qquad\qquad \Psi: R \to \hat{D},\ x = \Psi(Q),\ Q\varepsilon R,$$

où U est un ouvert quelconque de $\hat{D}$ contenu dans un secteur angulaire different de $\mathbb{C}$ et tel que $0 \in \bar{U}$.

La généralisation de cette théorie se fait en généralisant d'abord sa forme locale. Considérons

$$\mathcal{D} = \hat{D}^n \times D^{m-n} \subset \mathbb{C}^m, \quad \mathcal{R} = R^n \times D^{m-n}$$

et $\mathcal{H}^{p\times 1}(\mathcal{R})$ l'espace vectoriel sur $\mathbb{C}$ des applications holomorphes de $\mathcal{R}$ dans $\mathbb{C}^p$. Si E est un sous-espace vectoriel de dimension finie q de $\mathcal{H}^{p\times 1}(\mathcal{R})$ possèdant les deux propriétés suivantes:

A) E est stable par les opérations de $\pi_1(\mathcal{D})$

B) tous les éléments de E sont d'ordre fini à l'origine (généralisation de (2) en utilisant des ouverts convenables de $\mathcal{D}$)

alors E admet une base de la forme:

$$(3) \qquad (e_1, e_2, \ldots, e_q) = V(M) \prod_{j=1}^{n} Q_j^{A_j} \times \prod_{j=1}^{n} Q_j^{F_j}$$

$$(Q = (Q_1, Q_2, \ldots, Q_n) \in \mathcal{R},\ M = \Psi(Q) \in D^m)$$

où les matrices A_j (resp. F_j) ont en particuliers les propriétés suivantes: elles sont constantes et diagonales (resp. nilpotentes sous forme triangulaires); et où V est une matrice holomorphe à l'origine.

Si le rang de V(0) est maximum alors E est dit faiblement singulier à l'origine. Tout sous-espace vectoriel de dimension finie p de $\mathcal{H}^{p\times 1}(\mathcal{R})$ faiblement singulier à l'origine est l'espace vectoriel des solutions d'une système de la forme

$$(4) \qquad df = \left(\sum_{j=1}^{n} \frac{P_j}{x_j} dx_j + \sum_{j=n+1}^{m} P_j\, dx_j\right) f$$

où les P_j sont méromorphes dans D^m et holomorphes à l'origine. Réciproquement l'espace vectoriel des solutions d'une système de

Pfaff (4) complètement intégrable est faiblement singulier à l'origine de $\mathbb{C}^m$.

Soient V une variété analytique complexe de dimension m, A un sous-ensemble analytique de V de codimension un en chacun de ses points. Si les composantes irréductibles de A sont en position générale et sans singularité il est facile de définir la notion d'espace vectoriel faiblement singulier sur V. Un système de Pfaff complètement intégrable

$$(s) \qquad df = \omega f$$

où ω est holomorphe sur V-A est dit de Fuchs si l'espace vectoriel de ses solutions est faiblement singulier sur V. La théorie locale conduit à définir une classe $\Omega^{p\times p}$ (V,A) de formes différentielles à valeurs matricielles et à montrer:

(F) Un système de Pfaff complètement intégrable (s) est de Fuchs si et seulement si $\omega \in \Omega^{p\times p}(V,A)$.

Lorsque les composantes irréductibles de A ne sont plus en position générale nous savons qu'il existe une modification de Hopf nous ramenant au cas précèdant. En utilisant le fait que la classe $\Omega^{p\times p}(V,A)$ (qui peut être définie même si A possede des singularités) est stable par modification de Hopf; on démontre le résultat (F) dans le cas général.

Les applications de cette théorie sont analogue à celles de la théorie classique de Fuchs et donnent des systèmes de Pfaff relativement simples, associés aux fonctions hypergéometriques.

Les détails de ce travail sont exposés dans [1].

RÉFÉRENCES:

[1] Gérard, R. Théorie de Fuchs sur une variété analytique complexe, J.de Math. Pures et Appliquées 47 (1968) 135-152.

(50) Invariant subsets of hyperbolic sets M. Hirsch

Let Λ be an invariant set of a C^{∞} diffeomorphism $f : M \to M$; then Λ is hyperbolic if $TM = E^u \oplus E^s$ where E^u, E^s are invariant under Tf and there exist $c > 0$, $\lambda > 1$ such that $|Tf^n x| \geq c\lambda^n |x|$ $(x \in E^u, n \geq 0)$ and $|Tf^{-n} x| \geq c\lambda^n |x|$ $(x \in E^s, n \geq 0)$. If $\Lambda = M$ then f is Anosov. Since $F|\Lambda$ is expansive it is known, for example, that if $X \subset \Lambda$ is a compact invariant set of f then $X \not\approx S^1$.

Let $f: T^n \to T^n$ be a hyperbolic toral automorphism, i.e. f is induced by $\bar{f} : R^n \to R^n$ given by an integer $n \times n$ matrix with $\det \bar{f} = \pm 1$ and no eigenvalues on the unit circle. Let X denote a compact, locally connected invariant set of f.

THEOREM 1. If the rank of the image of $H_1(X) \to H_1(T)$ is ≤ 1, then X is a finite union of periodic orbits.

Corollary. Invariant subsets cannot be homeomorphic to $S^k (k>0)$, RP^k, CP^k, Klein bottle, etc.

THEOREM 2. If X is connected and contains a periodic point then either X = one point or there is an f-invariant toral subgroup $G \subset T^n$ with $X \subset G$ and $\pi_1(X) \to \pi_1(G)$ surjective.

THEOREM 3. If $X = M^k$ (a topological submanifold of T^n), M^k contains a periodic point (always true if $k \leq 2$) and rank $(H_1(M^k) \to H_1(T^n)) = r$, then (a) $r \geq k$, and (b) $r = k \Rightarrow M^k$ is a toral subgroup. (Proof immediate from Theorem 2).

Problem. If $n = 4$ can $k = 2$ (genus $M^2 \geq 2$)? (As we see later, M^2 cannot be a smooth submanifold).

THEOREM 4. If X is connected, and the stable manifolds are 1-dimensional, then X = one point (or $X = T^n$).

THEOREM 5. $X \neq T^n \Rightarrow \dim X \leq n - 2$ (X not necessarily connected or locally connected).

Problem. Is $\dim X = 1$ possible?

Now we return to a general $f : M \to M$; let V be a compact invariant smooth submanifold which is also a hyperbolic set.

Conjecture. $f|V : V \to V$ is Anosov.

THEOREM 6 (Hirsch and Pugh). Let Ω = nonwandering set for $f|V$. Then Ω is a hyperbolic set for $f|V$, and the splitting of TV over Ω is $(E^u \cap TV) \oplus (E^s \cap TV)$.

THEOREM 7. $f|V$ is Anosov in the following cases:

(a) $\dim E^u = 1$ or $\dim E^s = 1$

(b) $\dim E^u = 2$, some $x \in M$ has a dense forward orbit $O_+(x)$ (or similar statement with E^s, $O_-(x)$)

(c) codim $V = 2$, some x has dense $O_+(x)$ and some y has a dense $O_-(y)$

(d) codim $V = 1$, some x has dense orbit

(e) $\dim V = 2$.

Thus for a hyperbolic toral automorphism f, the first possibility for $f|V$ not being Anosov is $V^3 \subset T^6$, dim (stable manifolds) = 3, rank $(H_1(V) \to H_1(T)) = 6$.

R.F. Williams has proved that if $X \subset \Lambda$ is an invariant 1-dimensional continuum then X is not chainable.

Proofs will appear in the Proceedings of a Conference in Honour of G. de Rham (Springer).

(51) The principle of Maupertuis — G. Godbillon

The principle of Maupertuis states that the trajectories with constant energy of a Hamiltonian system are, perhaps with a change of time, the geodesics of a Riemannian structure on the configuration space.

Considered classically as a variational principle, it can be also put into a geometrical framework. See [1].

1 Hamiltonian flows.

Let M^{2n} be a smooth manifold with a symplectic structure, i.e. a closed 2-form ω of maximal class ($\omega^n \neq 0$ everywhere). A Hamiltonian flow on (M, ω) is a vector field X on M such that $\alpha = i_X\omega$ is a closed Pfaffian form. Conversely given ω closed there is a unique X with $\alpha = i_X\omega$.

Examples. Let V^n be a smooth manifold and $M^{2n} = T^*(V)$ be the cotangent space of V with projection $q : T^*(V) \to V$. If λ is the Pfaffian form on M defined by

$$<u, \lambda(\alpha)> = <q^T u, \alpha>, \quad \alpha \in T^*(V), \quad u \in T_\alpha(T^*(V)),$$

then $\omega = d\lambda$ is the symplectic form of Cartan on M; locally with respect to coordinates $(q_i, p_i = \frac{\partial}{\partial q_i})$ on M we have $\lambda = \Sigma p_i dq_i$ and $\omega = \Sigma dp_i \wedge dq_i$. Examples of Hamiltonian flows on $(T^*(V), \omega)$ are:

(i) geodesic flow given by $i_X\omega = -dT$ where $T:T^*(V) \to R$ is a Riemannian metric;

(ii) classical Hamiltonian flow given by $i_X\omega = -dH$ where $H = T-U\circ q$, $T:T^*(V) \to R$ Riemannian metric and $U:V \to R$.

The form $\alpha = i_X\omega$ defines a foliation of codimension 1 on the open set $U = \{x \in M | \alpha(x) \neq 0\}$, and X is tangent to the leaves. If N^{2n-1} is a leaf (with inclusion $i:N \to M$) then $i^*\omega$ is closed and $(i^*\omega)^{n-1} \neq 0$. When $\omega = d\lambda$ we have $di^*\lambda = i^*\omega$; so $i^*\lambda$ is of class $\geq 2n-2$. If moreover $\lambda(X) \neq 0$ on $i(N)$ then $i^*\lambda$ is of maximal class $2n-1$ on N^{2n-1} ($i^*\lambda \wedge (di^*\lambda)^{n-1} \neq 0$) : $i^*\lambda$ is a contact form.

2 Contact forms.

A contact form on a smooth manifold W^{2n+1} is a Pfaffian form ν of maximal class : $\nu \wedge (d\nu)^m \neq 0$. Locally a contact form is of type $\nu = dx_1 + x_2 dx_3 + \ldots + x_{2m} dx_{2m+1}$ (Darboux), so there is a unique vector field Z (locally $\frac{\partial}{\partial x_i}$) with $\nu(Z) = 1$ and $i_Z d\nu = 0$.

THEOREM (Reeb [3]). The vector field Z corresponding to a contact form ν on W^{2m+1} cannot have a closed section.

Proof. If S is a (connected) closed section of Z then $(d\nu)^m \neq 0$ on S (transversality); so $\int_S (d\nu)^m \neq 0$. But by Stokes' formula

$$\int_S (d\nu)^m = \int_{\partial S} \nu \wedge (d\nu)^{m-1} = 0 \quad (\partial S = \emptyset).$$

Problem (Chern): Does there exist a contact form on a compact orientable 3-manifold?

If c_t is a continuous family of "contact structures" (i.e. of hyperplane distributions which can be defined by contact forms), there exists an isotopy $h_t : W \to W$ such that $c_t = h_t^* c_0$ (Martinet, [2]).

Problem: Are all contact structures on S^3 (or W^{2m+1}) conjugate?

3 Principle of Maupertuis.

In example (i) above X is locally $\Sigma\left(\frac{\partial T}{\partial p_i}\frac{\partial}{\partial q_i} - \frac{\partial T}{\partial q_i}\frac{\partial}{\partial p_i}\right)$, and $\lambda(X) = \Sigma \frac{\partial T}{\partial p_i} p_i = 2T$. Hence λ induces a contact form on $N = T^{-1}(h)$, $h > 0$, with corresponding vector field $\frac{X}{2h}$.

In (ii) we have $X = \Sigma\left(\frac{\partial H}{\partial p_i}\frac{\partial}{\partial q_i} - \frac{\partial H}{\partial q_i}\frac{\partial}{\partial p_i}\right)$, and $\lambda(X) = \Sigma \frac{\partial H}{\partial p_i} p_i = 2T$. Hence λ induces a contact form on the manifold $\tilde{N} = \{x \in T^*(V) \mid H(x) = h, T(x) \neq 0\}$ (h a regular value of H), with corresponding vector field $Z = \frac{X}{2T}$.

Now $T' = \frac{T}{U+h}$ is a Riemannian metric on $T^*(W)$ where $W = \{x \in V \mid U(x) + h \neq 0\}$, and $\tilde{N} = T'^{-1}(1) \subset T^*(W)$. Thus the geodesic flow Y of T' given by $i_X\omega = -dT'$ is tangent to $\tilde{N}$, and λ induces a contact form on $\tilde{N}$ with corresponding vector field $Z = \frac{Y}{2T'}$. So we have $\frac{X}{2T} = \frac{Y}{2T'}$, or $X = (U + h)\, Y$; which is the principle of Maupertuis.

REFERENCES:

[1] Godbillon, C. Géometrie Différentielle et Mécanique Analytique, Hermann, Paris, 1969.

[2] Martinet, J. Thèse, Grenoble 1969.

[3] Reeb, G. Sur certaines propriétés topologiques des trajectoires des systèmes dynamiques, Acad. Roy. Belg. Cl. Sci. Mem. Coll. 8° 27 (1952), no. 9.

(52) Instability in $\mathrm{Diff}^r(T^3)$

C. Simon

A principal goal of Smale's program [2] has been to find and classify a Baire set D of diffeomorphisms on a given compact manifold M, such that all maps in D have sufficiently strong stability properties and can be classified by a countable and practical set of invariants, e.g. zeta functions. Among the candidates for D have been the Morse-Smale maps, the structurally stable maps, and diffeomorphisms satisfying Axiom A. Each new conjecture has followed from careful analysis of past counterexamples. Most recently, Abraham, Smale, Shub and Newhouse have exhibited counterexamples to the density of Smale's Axiom A and of Ω-stability. However, earlier in this conference, Abraham insisted that more such counterexamples must be constructed and understood for the theory to advance. Here we construct, the first C^1 counterexample on a 3-manifold, a map exhibiting some strong instability properties.

THEOREM. Let $r \geq 1$ and $\lambda >> 3$, λ an eigenvalue of a matrix in $SL(2,Z)$. If $f \in \mathrm{Diff}^r(T^3)$ let $N_n(f)$ be the number of fixed points of f^n. There exists an open set U in $\mathrm{Diff}^r(T^3)$ such that for any f in U, (1) f does not satisfy Smale's Axiom A, (2) f is not Ω-stable, (3) $\Omega(f)$ has a basic component Ω_1 that has both zero- and one-dimensional path components, (4) for any neighborhood V of f in U, $\exists$ f' ε V and an integer m such that $N_m(f) \neq N_m(f')$, (5) the zeta function is not constant on any open subset of U, (6) the zeta function is irrational for a Baire set of f in U, (7) log $\lambda \leq$ topological entropy of f $\leq$ log λ + log 3, (8) there is a dense set of f in U with non-hyperbolic periodic points.

<u>Lemma</u>: Let $A:T^2 \to T^2$ be a hyperbolic toral automorphism [2]. Then there exists $f:T^2 \to T^2$ smoothly isotopic to A such that a) f satisfies Axiom A; b) $\Omega(f) = \{\theta\} \cup \Lambda$ where $\{\theta\}$ is a point source (the origin in $T^2 = R^2/Z^2$) and Λ is a one-dimensional attractor, a "solenoid"; c) $\Lambda = \cup\{W^u(x) : x \in \Lambda\}$; d) for each x in Λ, $W^s(x)$ is dense in T^2; and e) f respects the foliation $\mathfrak{F}$ whose leaves are the stable manifolds for A.

<u>Proof</u>. f is the diffeomorphism DA of [1] or [2,I.9]. One merely alters A in a neighbourhood of θ by changing θ from a saddle-like fixed point to a source while adding two new saddle fixed points, x_0 and y_0.

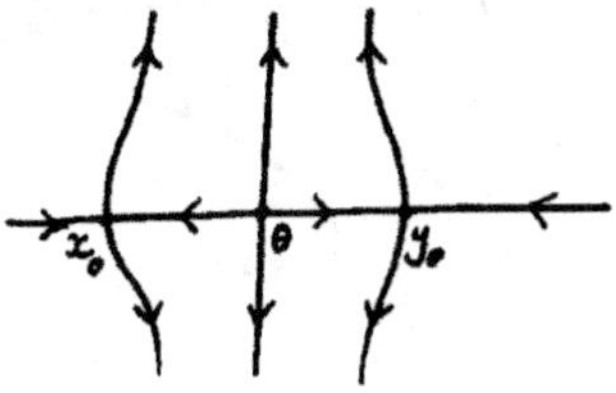

<u>CONSTRUCTION OF U</u>:

Let $g:S^1 \to S^1$ be a hyperbolic diffeomorphism with one sink, {-1}, and one source, {+1}. Consider $f \times g : T^2 \times S^1 \to T^2 \times S^1$, f as in the lemma with one eigenvalue of A equal to λ of the hypothesis of our theorem. The map $f \times g$ preserves the foliation $\mathcal{G}$ whose leaves are $F \times S^1$, $F \in \mathfrak{F}$, and $\Omega(f \times g) = [\Lambda \times \{+1\}] \cup \{(\theta,+1)\} \cup [\Lambda \times \{-1\}] \cup \{(\theta,-1)\}$. We will now alter the first of these components. Let $\bar{x}_0$ denote the fixed point $(x_0,+1) \in T^2 \times S^1$, T_1^2 denote $T^2 \times \{+1\}$, and Λ_1 denote $\Lambda \times \{+1\}$.

Now $W^u(\bar{x}_0)$ is a 2-dimensional "wall" transversal to T_1^2.

We want to force $W^u(\bar{x}_o)$ back through T_1^2. To do this, choose an open set N in T^3 shaped as pictured where $N \cap T_1^2$ is a 2-disk B_1 and $W^u(\bar{x}_o)$ meets N for the first time at the 2-disk B_2. Choose N such that $(f \times g)B_i \cap B_i = \emptyset$, $i = 1,2$. Let $b:T^3 \to T^3$ be a C^r bump function which 1) is the identity outside N, 2) maps some points of the interior of B_2 on and below B_1, and 3) respects the foliation $\mathcal{G}$. Then, $h = b \circ (f \times g)$ has a flow picture which somewhat simplified looks like figure A in T^3 and figure B in T_1^2.

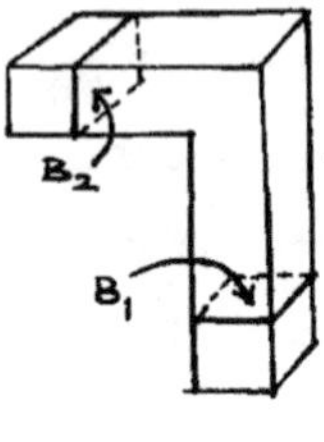

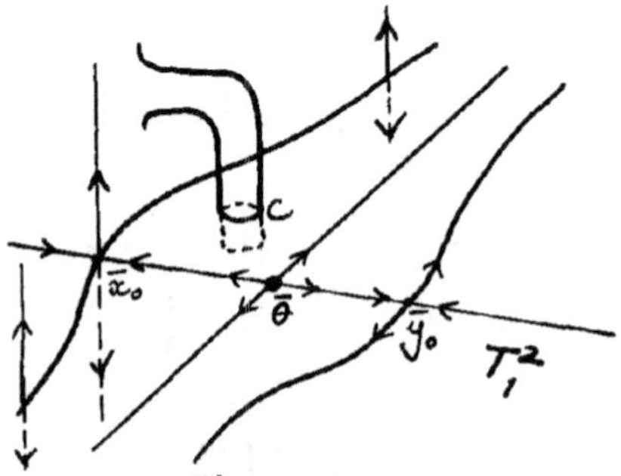

Figure A

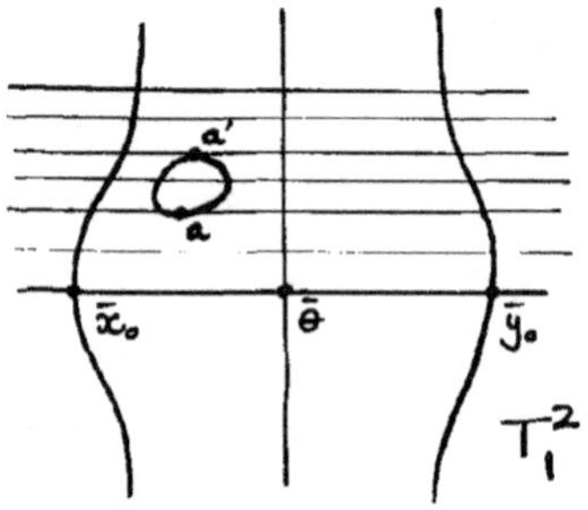

Figure B

Let c be the circle where $b(B_2)$ intersects B_1 in T_1^2. Since $W^s(\Lambda_1, f \times g) \cap B_2 = W^s(\Lambda_1, h) \cap B_2$, we have $c \subset \Omega(h)$, using [2,I.7]. But at least two points of c are tangent to leaves of $\mathcal{F}$, say a and a'. Because of this tangency, non-wandering points a and a' cannot be hyperbolic. One notes that these phenomena occur for small perturbations of h; thus we define U to be a small neighbourhood of h. In fact, by a recent theorem of Hirsch, Pugh, and Shub, because h is more hyperbolic normal to leaves of $\mathcal{G}$ than on leaves of $\mathcal{G}$ ($\lambda >> 3$), any perturbation of h respects a new foliation of T^3 which is conjugate to $\mathcal{G}$. In this example, this is a result of the structural stability of Λ_1. Thus, one analyzes perturbations of h just as one analyzes h. It is easy to see that periodic points of h lie only on leaves which contain a periodic point in Λ_1. In fact, $h^n|\Omega(h^n|(\text{leaf of period } n))$ is conjugate to the shift map acting on a quotient space of $(3^n)^Z$, the bisequence set on 3^n symbols. By perturbing h near the point a, one can alter the number of periodic points on periodic leaves near a, obtaining (4). (2) and (5) follow from (4). (6) uses (4), the countability of the set of rational zeta functions, and a Baire category argument. (8) follows from a careful analysis of the changes occurring near c. The rationale for (7) is that $\Omega(h)$ is more than the solenoid Λ and less than $\Lambda \times \{3\text{-shift}\}$.

REFERENCES:

[1] Shub, M. Instability, these _Proceedings_.

[2] Smale, S. Differentiable dynamical systems, _Bull. A.M.S._ 73 (1967) 747-817.

(53) _A global concept of stability under persistent perturbations_ — _P. Seibert_

1. Let X be a compact differentiable manifold with a metric. We consider the flow (X,π) defined by the differential equation $\dot{x} = f(x)$ ($x \in X$, f Lipschitz). Define a _δ-solution_ ($\delta > 0$) to mean an absolutely continuous function $\sigma : R \to X$ such that $||\dot{\sigma}(t) - f(\sigma(t))|| < \delta$ a.e.. Let $P_\delta(x,t) = \{y | \exists$ a δ-solution σ with $\sigma(0) = x, \sigma(t) = y\}$, $P_\delta^\pm(x) = P_\delta(x, R^\pm)$ ($R^+ = [0,\infty)$, $R^- = (-\infty,0]$), $P^\pm(x) = \bigcap_{\delta>0} P_\delta^\pm(x)$ (consisting of the points attainable from x by δ-solutions with δ arbitrarily small). A compact set M is called _totally stable_ [_negatively t.s._] if given $\varepsilon > 0$, there exists $\delta > 0$ such that $P_\delta^+(M)$ $[P_\delta^-(M)] \subset$ an ε-neighbourhood of M (equivalent to the definition of Dubošin (1940), Malkin (1944); cf. [3], §28; [2],

§24).

Properties of $P^{\pm}$: (i) $x_n \to x$, $y_n \to y$, $y_n \to P^{\pm}_{\delta_n}(x_n)$, $\delta_n \to 0$ $\Rightarrow$ $y \in P^{\pm}(x)$, [2]. (ii) M compact $\Rightarrow$ $\dot{P}^{\pm}(M)$ compact [(i) $\Rightarrow$ (ii)]. (iii) $P^{\pm}$ are transitive, [2]. (iv) If M compact, then M totally stable $\Leftrightarrow$ $P^{+}(M) = M$, [2]. (v) $P^{+}(M)$ compact $\Rightarrow$ $P^{+}(M)$ totally stable [(iii) and (iv) $\Rightarrow$ (v)].

2. Define A to be <u>contracting</u> if $t > 0 \Rightarrow \bar{A}t \subset \text{Int } A$, <u>expanding</u> if its complement is contracting. [Notation: $\pi(x,t) = xt$].

<u>THEOREM 1</u> [4]. A compact set M is totally stable $\Leftrightarrow$ M has a fundamental system of contracting neighbourhoods.

Thus total stability is a purely topological concept, and we can define totally stable sets for continuous flows.

<u>THEOREM 1'</u> [4]. Compact M is totally stable $\Leftrightarrow$ M has a fundamental system of asymptotically stable neighbourhoods.

<u>THEOREM 2</u>. $P^{\pm}(x) = \bigcap \{$all contracting (expanding) closed sets containing $x\}$. (Consequence of [2] and [4].)

In the case of a continuous flow, we may use this property to define $P^{\pm}(x)$.

3. Let $D^{\pm}(x) = \{y \mid \exists\ x_n \to x,\ t_n \in R^{\pm}$ such that $x_n t_n \to y\}$ (the "prolongations" of x (Ura)). Define a continuous flow to be <u>totally stable</u> if $P^{\pm}(x) = D^{\pm}(x)$, for all x. (In any case $D^{\pm}(x) \subset P^{\pm}(x)$ because of (i).) In other words, small initial perturbations have the same effect on a given motion as small perturbations of the system.

<u>THEOREM 3</u>. A continuous flow on a compact phase space is totally stable $\Leftrightarrow$ (a) all compact (Liapunov) stable sets are totally stable, and (b) $D^{\pm}$ are transitive.

Example where (a) fails: flow with centre. - Examples where (b) fails: 1. flow with two linked saddles; 2. flow with semi-stable limit cycle; 3. 1-point compactification of parallel flow.

REFERENCES:

[1] Auslander, J., N.P. Bhatia and P. Seibert — Attractors in dynamical systems, Bol. Soc. Mat. Mexicana 9 (1964) 55-66.

[2] Auslander, J. and P. Seibert — Prolongations and stability in dynamical systems, Ann. Inst. Fourier, Grenoble 14 (1964) 237-268.

[3] Hahn, W. — Theory and Applications of Liapunov's Direct Method, Prentice-Hall, 1963.

[4] Seibert, P. — Estabilidad bajo perturbaciones sostenidas y su generalización en flujos continuos, Acta Mex. de Ciencia y Tecnol. 2 (1968) 154-165; correction to appear.

(54) Hausdorff dimension and transversality of discrete flows

H. Furstenberg

Ergodic theory may be useful in attacking problems in diophantine approximation, such as finding the distribution of the fractional parts of powers of a given rational.

Specifically, let $r = p/q$ be a rational > 1 and form the discrete subgroup of rationals $\Gamma = \bigcup_{n=1}^{\infty} Z/p^n q^n$, Z the integers. The dual $X = \hat{\Gamma}$ is a compact group and each automorphism of Γ induces an automorphism of X. In particular the operations of multiplying by pq and by p/q generate groups S_1 and S_2 of automorphisms of X. It is not hard to show that there is an element $x_o \in X$ with the property that the orbit $S_2 x_o$ being dense in X implies that the sequence r^n is dense modulo 1. What can be shown is that for all non-torsion elements $x \in X$ (including x_o), the orbit $S_1 S_2 x$ is dense in X. Moreover $\overline{S_1 x_o}$ is a countable set. This suggests in this context:

Conjecture. S_1, S_2 commuting semigroups $\Rightarrow$ dim $\overline{S_1(x)}$ + dim $\overline{S_2(x)}$

$\geq$ dim X for all non-torsion $x \in X$, for some suitable definition of dimension.

A consequence of this would be that whenever dim $\overline{S_1(x)} = 0$, dim $\overline{S_2(x)}$ = dim X. In the situation in question this is known to imply $\overline{S_2(x)} = X$ and the above conjecture would at least contribute to the equidistribution problem for $S_2(x)$. We analyze more explicitly the following analogous case.

If $X = S^1 = R/Z$, $p,q \in Z$ (>0) not both powers of the same integer, $S_1(S_2)$ = {endomorphisms generated by $x \mapsto px(qx) \bmod Z$}, then we have

<u>THEOREM 1</u>. x irrational $\Rightarrow \overline{S_1 S_2(x)}$ dense in X.

Let A(B) be an $S_1(S_2)$-invariant closed set. Then the above conjecture implies dim A + dim B < 1 $\Rightarrow A \cap B \subset$ rationals. Define two arbitrary closed sets C,D to be <u>transverse</u> if dim $C \cap D$ $\leq$ max {dim C + dim D - dim X, 0}.

<u>Conjecture</u>. S_1, S_2, A, B as above $\Rightarrow$ A + t, B transverse for all $t \in S^1$. In particular this implies $\overline{S_1 S_2(x)}$ transverse to itself, so dim $\overline{S_1 S_2(x)}$ = 1 or 0. One shows readily that the former implies $\overline{S_1 S_2(x)} = X$, the latter $\overline{S_1 S_2(x)} \subset$ rationals.

We study these problems using sequence spaces. Let $\Lambda = \{a_1, \ldots, a_n\}$, Ω_Λ = {sequence $\omega = (a_{i_1}, a_{i_2}, \ldots)$}, Λ^* = {words of finite length in the a_i, including 'zero' word 1}; then Λ^* is a semigroup under juxtaposition. A <u>tree</u> is a subset Δ of Λ^* satisfying $1 \in \Delta$ and $a_{i_1} \ldots a_{i_s} \in \Delta \Rightarrow a_{i_1} \ldots a_{i_r} \in \Delta$, $1 \leq r \leq s$. For example $\Lambda = \{a,b\}$,

Δ = { 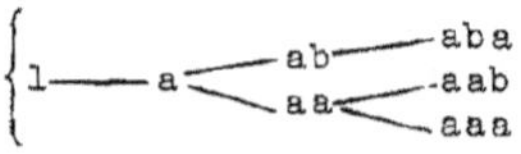 etc. }.

Trees over Λ correspond 1:1 to closed subsets of Ω_Λ (with product topology: $\Omega_\Lambda = \Lambda \times \Lambda \times \ldots$). A <u>cut</u> of a tree Δ is a subset of Δ

meeting each set of the form $\{\omega_j | \omega_1 = 1,\ \omega_{j+1} = \omega_j a_{i_j}\ (a_{i_j} \in \Lambda)\}$ exactly once. We now define the dimension of Δ by $\dim \Delta = \inf \{\lambda | \exists$ cuts containing arbitrarily long words, with $\sum_{\sigma \in \text{cut}} e^{-\lambda \ell(\sigma)} < 1$, $\ell(\sigma) = $ length of σ, $\lambda \in R\}$.

If $\Lambda = \{0, \ldots, r-1\}$ define $\phi : \Omega_\Lambda \to [0,1]$ by $(a_{i_1} a_{i_2} \ldots) \mapsto \sum_{n=1}^{\infty} (a_{i_n}/r^n)$.

Lemma. If $A(\Delta) \subset \Omega_\Lambda$ corresponds to the tree Δ then the Hausdorff dimension of $\phi A(\Delta) = \dim \Delta / \log r$.

Define arbitrary sets $A, B \subset R$ to be strongly transverse if $\dim (A + t) \cap B \leq \max \{\dim A + \dim B - 1, 0\}$, all $t \in R$.

THEOREM 2. A, B compact $\Rightarrow \dim (A+uB) \geq \min \{\dim A + \dim B, 1\}$ for almost all u.

Possible Theorem. A, B compact $\Rightarrow$ A, uB strongly transverse for almost all u.

THEOREM 3. Let $X = S^1$; S_1, S_2 as before, and let $A(B)$ be an $S_1(S_2)$-invariant set. Then if the 'possible theorem' is true, A,B are strongly transverse.

This follows from

THEOREM 4. A, B (as above) not strongly transverse $\Rightarrow$ for almost all u, uA and uB are not strongly transverse.

The proof uses the following main lemma:

Lemma. If R is a closed invariant set of trees (i.e. if $\Delta \in R$ then the 'successor' of Δ obtained by removing the first letter of every word in Δ also $\in R$) containing a tree of $\dim > \delta$, then there is an ergodic R-valued stationary process $\{z_n\}$ (i.e. $z : \Omega \to R$, $z_n(\omega) = z(T^n \omega)$, $\omega \in$ some (Ω, μ), T an ergodic m.p.t. of Ω) such that $\mu \{\omega | \dim z_n(\omega) \leq \delta\} = 0$.

Trees arise in Theorem 4 as follows: suppose $1 < p < q$, and let $W(\subset R^2) = \{(a,b;\ell)\,|\,(a,b) \in A \times B,\ \ell = \text{line through } (a,b) \text{ of slope } u,\ 1 < u < q\}$, where we consider S^1 as $[0,1] \subset R$. Define $T:W \to W$ by $T(x,y;\ell) = (S_1x, S_2y;\ell')$ (slope $(\ell') = (\frac{q}{p})u$) if $u < p$, and $T(x,y;\ell) = (S_1x,y;\ell'')$ (slope $(\ell'') = (\frac{1}{p})u$) if $u > p$. Thus we obtain a tree

$$\Delta(\ell) : \ell <^{\ell\ell' <}_{\ell\ell'' <} \quad \text{etc..}$$

Now in fact $\dim \Delta(\ell) = \text{Hausdorff dim } (\ell \cap A \times B) \times \log p$, and an application of the lemma shows that if one line meets $A \times B$ in a set of dimension $> \delta$ then in almost every direction there is a line meeting $A \times B$ in a set of dimension $> \delta$. This then implies Theorem 4.

(55) Universal foliations — A. Phillips

This seminar reports on recent work of Haefliger.

A foliation $\mathfrak{F}$ of codimension q on a topological space X means a covering $\{U_i\}$ of X by open sets, with maps $\{f_i : U_i \to R^q\}$ locally related by diffeomorphisms $\{\gamma_{ij}\}$ ($f_i = \gamma_{ij} \circ f_j$ locally) satisfying $\gamma_{ik} = \gamma_{ij}\gamma_{jk}$ where defined. If X is a manifold and the f_i are submersions then $\mathfrak{F}$ is regular. The normal bundle $\nu_{\mathfrak{F}}$ is $\bigcup_i f_i^* TR^q/\sim$, where $(x,v)\sim(y,w) \Leftrightarrow x = y,\ w = d\gamma_{ij}(f_j x).v$. Thus for example if $X = S^1 = \{e^{i\theta}\}$, $U_1 = \{\frac{-3\pi}{4} < \theta < \frac{3\pi}{4}\}$, $U_2 = \{\frac{\pi}{4} < \theta < \frac{7\pi}{4}\}$, $f_1(\theta) = |\theta|$, $f_2(\theta) = \theta$, then $\gamma_{12}(t) = t$ on one component of $U_1 \cap U_2$, $= 2\pi - t$ on the other, and we have a foliation $\mathfrak{F}$ with $\nu_{\mathfrak{F}}$ = nontrivial line bundle over S^1. If $\mathfrak{F}$ is defined by a unique submersion $f : M \to R^q$ then the bundle $(\ker df)^\perp \subset TM$ (with respect to some metric) is isomorphic to $\nu_{\mathfrak{F}}$. Similarly for an arbitrary regular foliation $\mathfrak{F}$ of M there is a well-defined 'kernel' bundle orthogonal to $\nu_{\mathfrak{F}} \subset TM$.

Any continuous $h:Y \to X$ induces $h^*\mathfrak{F}$ on Y from $\mathfrak{F}$ on X, and it is easy to check that $\nu_{h^*\mathfrak{F}} = h^*\nu_{\mathfrak{F}}$. If X,Y are manifolds and $\mathfrak{F}$ is regular, then each $f_i \circ h$ a submersion (i.e. h transverse to $\mathfrak{F}$) $\Rightarrow h^*\mathfrak{F}$ regular.

Foliations $\mathfrak{F}_0, \mathfrak{F}_1$ (codim q) on X are homotopic ($\mathfrak{F}_0 \sim \mathfrak{F}_1$) if there is a foliation $\mathfrak{H}$ (codim q) on $X \times I$ inducing $\mathfrak{F}_j$ on $X \times \{j\}$, and the same as $p_j^*\mathfrak{F}_j$ in a neighbourhood of $X \times \{j\}$ (p_j = projection $X \times I \to X \times \{j\}$), $j = 0,1$. Clearly $f \simeq g : Y \to X$ $\Rightarrow f^*\mathfrak{F} \sim g^*\mathfrak{F}$, and also $\mathfrak{F}_0 \sim \mathfrak{F}_1 \Rightarrow \nu_{\mathfrak{F}_0} = \nu_{\mathfrak{F}_1}$. It is easy to show $\mathfrak{F}_0, \mathfrak{F}_1$ regular, codim n on $M^n \Rightarrow \mathfrak{F}_0 \sim \mathfrak{F}_1$.

Define the functor F_q : {CW complexes, homotopy classes of maps} $\to$ {sets} by $F_q(X)$ = (homotopy classes of foliations of codim q on X). This is a 'homotopy functor' (Brown), so

Proposition. There is a CW complex $B\mathfrak{F}_q$ such that $F_q(X) = [X, B\mathfrak{F}_q]$.

Corresponding to id : $B\mathfrak{F}_q \to B\mathfrak{F}_q$ is the universal foliation $\mathcal{E}$ on $B\mathfrak{F}_q$.

A q-dimensional foliated microbundle over X is a space E with maps $X \overset{i}{\to} E \overset{p}{\to} X$ ($pi = 1$) and a covering $\{U_i\}$ of X such that there are homeomorphisms $h_i : p^{-1}U_i \to U_i \times R^q$ with $h_j h_i^{-1}(x,t) = (x, \gamma_{ji}(t))$ (γ_{ji} diffeomorphism). Such E has a canonical foliation $\mathfrak{F}_E$, with cover $\{p^{-1}U_i\}$ and maps $p^{-1}U_i \overset{h_i}{\to} U_i \times R^q \to R^q$.

Proposition. Given $\mathfrak{F}$ on X there exists E over X such that $\mathfrak{F} = i^*\mathfrak{F}_E$.

THEOREM. M open manifold, $\mathfrak{F}$ on M $\Rightarrow \mathfrak{F} \sim$ regular foliation if and only if $\nu_{\mathfrak{F}}$ embeds in TM.

Lemma (Gromov; Phillips). M open $\Rightarrow$ the function {maps $M \to B\mathfrak{F}_q$ transverse to $\mathcal{E}$} $\to$ {epimorphisms $TM \to \nu_{\mathcal{E}}$} induced by differentiation is a weak homotopy equivalence.

THEOREM 2. If M is open and $\sigma \subset TM$ is a section of codimension q, then $\sigma \sim$ integrable section $\Leftrightarrow$ the bundle $\sigma^\perp$ can be induced from $\nu_{\mathcal{E}}$.

The bundle $\nu_{\mathcal{E}}$ over $B\mathcal{F}_q$ corresponds to a classifying map $\alpha : B\mathcal{F}_q \to BO_q$. Results of Bott imply that α is not a homotopy equivalence, although it is a q-equivalence ($\alpha_* : \pi_n(B\mathcal{F}_q) \to \pi_n(BO_q)$ is iso if $n < q$, epi if $n = q$).

(56) Foliations of the plane — C. Godbillon

A foliation $\mathcal{F}$ of the plane R^2 has the following well known properties:

i) it is orientable;

ii) each leaf of $\mathcal{F}$ is closed in R^2 and homeomorphic to R (Poincaré - Bendixson);

iii) the space of leaves $X = R^2/\mathcal{F}$ is a 1-dimensional simply connected manifold (Haefliger - Reeb [3]);

iv) the projection $p : R^2 \to X$ is a locally trivial fibration (Whitney).

One can think of 4 types of classifications for foliations of the plane:

A) conjugacy of oriented foliations by homeomorphisms preserving the orientation;

B) conjugacy of oriented foliations by homeomorphisms;

C) conjugacy of foliations by homeomorphisms preserving the orientation;

D) conjugacy of foliations by homeomorphisms.

In each of these cases two conjugate foliations have homeomorphic space of leaves.

Kaplan [4] showed that problem A can be translated into an algebraic context (which he called "chordal systems").

But by properties iii) and iv) these problems are also equivalent to the problems of classifying locally trivial fibrations with fibre R over 1-dimensional simply-connected manifolds : if η = (E, p, X) is such a fibration whose total space E is Hausdorff, then E is homeomorphic to R^2, and the fibres of η are leaves of a foliation of R^2 (with X as space of leaves). The orientation of the foliation corresponds to a reduction of the structural group of η to the subgroup of increasing homeomorphisms of R, and the orientation of R^2 to an ordering on X locally isomorphic with the ordering of R.

So we have to give:

- a classification of 1-dimensional simply connected manifolds;
- a description of the group of homeomorphisms of such a manifold;
- a classification of locally trivial fibrations with fibre R over such a manifold.

Example. If X is the "branchement simple" (viz: 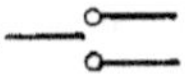) then there exist 2 classes of foliations (having X as space of leaves) in problem A, and 1 in each of the three other problems (Godbillon-Reeb [2]).

Using this result as a local model we can then reduce the above problem, in the case of manifolds having finite numbers of branching points, to (finite) combinatorial problems [1].

REFERENCES:

[1] Godbillon, C. (To appear).

[2] Godbillon, C. and G. Reeb. Fibrés sur le branchement simple, l'Enseignement Mathématique 12 (1966) 277-287.

[3] Haefliger, A. and G. Reeb. Variétés (non séparées) à une dimension et structures feuilletées du plan, l'Enseignement Mathématique 3 (1957) 107-125.

[4] Kaplan, W. Regular curve-families filling the plane II, Duke Math. J. 8 (1941) 11-46.

(57) Synthesis of control systems on manifolds — A. Halanay

We announce results obtained by Stefan Mirica in Bucharest. Let X be an n-dimensional C^1 manifold, Ω a p-dimensional manifold (possibly with boundary) called the space of controls; let $\xi : X \times \Omega \to T(X)$ be a parametrized vector-field, $\mathcal{J}$ a k-dimensional manifold (possibly with boundary) called the terminal manifold, and let $f^0 : X \times \Omega \to R$, $g : \mathcal{J} \to R$ be C^1 functions. Then $S = (X, \Omega, \xi, \mathcal{J}, f^0, g)$ defines a control system.

An admissible synthesis is defined by:

(a) a piecewise smooth set $N \subset X$ of dimension $\leq n - 1$

(b) piecewise smooth sets $P^k, P^{k+1}, \ldots, P^{n-1}$ with $\dim P^i = i$, such that $\mathcal{J} \subset P^k \subset \ldots \subset P^{n-1} \subset X$

(c) a function $v : X \to \Omega$, with the properties essentially required by Boltyanskii [3] and Mirica [6].

(In the definition of a piecewise smooth set P we require that any compact set in X intersect only a finite number of polyhedra in P).

Let σ be a cell, and define $\sigma' = \sigma$ if σ is of type I or $\sigma' = \sigma_0 \cup \sigma$ if σ is of type II and $\sigma_0 = \Sigma(\sigma)$ (see [3]). For each $x \in \sigma'$ a 'marked' trajectory $\phi_x(t)$ is well defined, and passes through a finite collection $\sigma_1, \ldots, \sigma_q$ of cells of type I, where $\pi(\sigma_q)$ is a connected component of $\mathcal{J}$. If this trajectory intersects $\pi(\sigma_i)$ in the point $\chi_j(x)$ at the moment $\tau_j(x)$, the functions τ_j and χ_j are proved to be C^1. Define

$$W(x) = g(\chi_q(x)) + \sum_{i=1}^{q} \int_{\tau_{i-1}}^{\tau_i} f^0(\phi_x(t), v(\phi_x(t)))\, dt,$$

and let $M = N \cup \left(\bigcup_{i=1}^{n-1} P^i\right)$; $X \backslash M$ is the union of cells of maximal dimension and for each $\sigma' \subset X \backslash M$ the restriction of W to σ' is a C^1 function.

THEOREM. If $\langle T_xW, \xi(x,\omega)\rangle + f^0(x,\omega) \geq 0$ for all $\omega \in \Omega$, $x \in X\setminus M$ then the marked trajectories of the corresponding admissible synthesis are optimal.

Let σ be a cell of type I, $\tilde{\xi}_\sigma : \tilde{\sigma} \to T(\tilde{\sigma})$ the field defined by $\tilde{\xi}_\sigma(x) = \xi(x,\tilde{v}_\sigma(x))$, where $\tilde{v}_\sigma : \tilde{\sigma} \to \Omega$ is a C^1 extension of v to an open neighbourhood $\tilde{\sigma} \subset X$ of $\bar{\sigma}$. Define $\tilde{f}^0_\sigma : \tilde{\sigma} \to R$ by $\tilde{f}^0_\sigma(x) = f^0(x,\tilde{v}_\sigma(x))$, $\tilde{H}_\sigma : T^*\tilde{\sigma} \to R$ by $\tilde{H}_\sigma(y) = \tilde{f}^0_\sigma(x) + \langle y, \tilde{\xi}_\sigma(x)\rangle$ for $y \in T^*_x\tilde{\sigma}$, $x \in \tilde{\sigma}$. The function $\tilde{H}_\sigma$ is C^1, and defines a field $\xi_{\tilde{H}_\sigma}$ on $T^*\tilde{\sigma}$ which is given locally by $\frac{d\alpha^i}{dt} = \tilde{f}^i_{\sigma,\alpha}(\alpha^1, \ldots, \alpha^n)$,

$$\frac{d\lambda_i}{dt} = \frac{\partial \tilde{f}^0_{\sigma,\alpha}}{\partial \alpha^i}(\alpha^1, \ldots, \alpha^n) - \Sigma\lambda_j \frac{\partial \tilde{f}^j_{\sigma,\alpha}}{\partial \alpha^i}(\alpha^1, \ldots, \alpha^n).$$

This construction can be made for each of the cells $\sigma_1, \ldots, \sigma_q$; denote $\tilde{H}_{\sigma_1}$ by $\tilde{H}_i$ and let $\xi_{\tilde{H}_i}$ be the corresponding field. Then it is proved that for each $x \in X\setminus M$ and each corresponding $\phi_x : [0,t_F] \to X$ there exist $y(x) \in T^*_xX$ and $\phi_{y(x)} : [0,t_F] \to T^*X$ such that $\phi_{y(x)}(0) = y(x)$, $\phi_{y(x)}(t) \in T^*_{\phi_x(t)}X$ for $t \in [0,t_F]$, on $[\tau_{i-1}(x), \tau_i(x)]$, $\phi_{y(x)}$ is an integral curve of $\xi_{\tilde{H}_i}$, $\phi_{y(x)}(\tau_i(x) - 0)$ ($= y_i^-$, say) and $\phi_{y(x)}(\tau_i(x) + 0)$ ($= y_i^+$) satisfy

$$y_q^- \circ T_{\chi_{q-1}(x)}\, i_{\tilde{\pi}(\sigma_q)} = T_{\chi_q(x)}\, g, \qquad \tilde{H}_q(y_q^-) = 0$$

$$y_r^- \circ T_{\chi_r(x)}\, i_{\tilde{\pi}(\sigma_r)} = y_r^+ \circ T_{\chi_r(x)}\, i_{\tilde{\pi}(\sigma_r)}, \qquad \tilde{H}_r(y_r^-) = 0,$$

$$r = 1, \ldots, q-1$$

($i_{\tilde{\pi}(\sigma_r)} : \tilde{\pi}(\sigma_r) \to \tilde{\sigma}_r$ being the inclusion mapping). Moreover, $T_xW = y(x)$, hence for the points in the cells of maximal dimension $y(x)$ is uniquely defined. Using this result the theorem may be restated in the form of the "maximum" principle (actually in this construction it will be a minimum).

We also describe an approach to the problem of constructing an optimal synthesis.

Further references for related questions are [1, 2, 4, 5].

REFERENCES:

[1] Albrecht, F. Control vector fields on manifolds and attainability, Lecture notes in operations research and mathematical economics, vol. 12, Springer, 1969.

[2] Berkovits, L.D. Necessary conditions for optimal strategies in a class of differential games and control problems, S.I.A.M. J. Control 5 (1967) 1-24.

[3] Boltyanskii, V.G. Sufficient conditions for optimality and the justification of the dynamic programming method, S.I.A.M. J. Control 4 (1966) 326-361.

[4] Isaacs, R. Differential games, J. Wiley and Sons, New York, 1964.

[5] Jones, G. Stephen and A. Strauss An example of optimal control, S.I.A.M. Review 10 (1968) 25-55.

[6] Mirica, S. On the admissible synthesis in optimal control theory and differential games, S.I.A.M. J.Control 7 (1969) 292-316.

(58) Foliations and transformation groups — M. Hirsch

We generalize the theorem that Anosov diffeomorphisms are structurally stable and have periodic points dense in the non-wandering set to the situation of a manifold M with foliation $\mathcal{L}$ and diffeomorphism f hyperbolic to $\mathcal{L}$. This means that f preserves $\mathcal{L}$, and TM splits into $E^u \oplus E^s \oplus L$ where L_x is tangent to the leaf $\mathcal{L}_x$ through x, each component is invariant under Tf, and Tf expands E^u and contracts E^s more than it expands or contracts L.

THEOREM 1. If f is hyperbolic to $\mathcal{L}$ there exists $\delta > 0$ such that if $g \in \mathrm{Diff}(M)$ is sufficiently (C^1) close to f then there is a unique g-invariant foliation $\mathcal{L}'$ of M with $|\mathcal{L} - \mathcal{L}'| < \delta$ and g hyperbolic to $\mathcal{L}'$, and a homeomorphism $h : M \to M$ such that

$h(\mathcal{L}_x) = \mathcal{L}'_{hx}$ and $hf\mathcal{L}_x = gh\mathcal{L}_x$.

A C^∞ action of a Lie group G on M is <u>locally free</u> if dim Gx = dim G for all x. Then the components of the orbits $\{Gx\}_{x \in M}$ form a foliation of M of dimension dim G. Define $g_0 \in G$ to be an <u>Anosov element</u> if g_0 is hyperbolic to this orbit foliation. If G contains an Anosov element, we have an <u>Anosov action</u>.

<u>THEOREM 2</u>. Anosov actions on a compact manifold are structurally stable.

The proof follows easily from Theorem 1.

<u>THEOREM 3</u>. Let F be a connected nilpotent Lie group. Let G act on compact M with $g_0 \in G$ an Anosov element, and let Ω = non-wandering set of a 1-parameter subgroup H containing g_0, considering H as a flow on M. Then given $x \in \Omega$ and $\varepsilon > 0$ there exists $y \in M$ with $d(y,x) < \varepsilon$ and Gy compact.

The assumption that G is nilpotent cannot be weakened to G solvable.

A non-trivial example of an Anosov action can be constructed taking $M = SL(3,R)/\Gamma$ where Γ is a uniform discrete subgroup, and $G = R^2$ = 3 × 3 diagonal matrices of determinant 1, acting on M by translation (L. W. Green).

For more details, see forthcoming articles in the Bulletin of the American Mathematical Society.

Many of the above results were obtained jointly with C.C. Pugh and M. Shub.

(59) Report on Bott's theorem on foliations J. Wood

Let M be a C^∞ manifold with tangent TM, and let [,] be the usual bracket operation on the sections ΓTM of TM (i.e. vector fields). A sub-bundle E of TM is integrable if ΓE is closed under [,], in which case we say E gives (or is) a foliation of M.

THEOREM (Frobenius' Theorem). E is integrable $\Leftrightarrow$ there are coordinate charts $(x^1, \ldots, x^m)$ such that the partial derivatives $\partial_{q+1}, \ldots, \partial_m$ span E.

The surfaces defined locally by x_1 = const., ..., x_q = const. are the leaves of the foliation.

There are two important questions that can be asked:

(1) Given a vector bundle $E \to M$, is E equivalent to $E' \subset TM$ which is integrable?

(2) Given $E \subset TM$, is E homotopic to an integrable $E' \subset TM$?

Until recently there were no known criteria for integrability, but in 1968 there was proved:

Bott's integrability criterion [1]. If $E \subset TM$ is integrable, then the subring Pont(TM/E) of the rational cohomology ring generated by Pontrjagin classes vanishes in dimension > 2 dim (TM/E).

Thus for example CP^5, CP^7, CP^9 etc. have no foliations of codimension 2, although TCP^5 etc. do have sub-bundles.

Note: A standard construction for foliations is to take a fibre bundle (or submersion) $f : M^m \to B^q$ and define E = ker (Df); the leaves are components of $f^{-1}(b)$, $b \in B$. Here $TM/E \simeq TB$ and Pont (TM/E) vanishes in dim > q.

Sketch proof of Bott's criterion. Let Q = TM/E, and define a partial connection $\nabla_X : \Gamma Q \to \Gamma Q$ by $\nabla_X S = \pi\left[X, \hat{S}\right]$ for $X \in \Gamma E$, $\pi\hat{S} = S$ where π is the projection $TM \to Q$. This is well-defined (by integrability), is a connection by the [,] rules, and the curvature tensor $K(X, Y)$ ($= [\nabla_X, \nabla_Y] - \nabla_{[X, Y]}$) is zero by the Jacobi identity. Using partitions of unity ∇_X can be extended to fields $X \in \Gamma TM$. Now locally $\partial_1, \ldots, \partial_q$ form a basis for Q, so $K(X, Y) : \Gamma Q \to \Gamma Q$ is given by a $q \times q$ matrix Ω^a_b of 2-forms on M (if $S = \sum_{b=1}^{q} s^b \partial_b$ then $K(X, Y)S = \sum_{a,b=1}^{q} \Omega^a_b (X, Y) s^b \partial_a$). Let $\Omega^a_b = \sum_{1 \leq i < j \leq m} R^a_{bij} \, dx^i \wedge dx^j$. Since $K(X, Y) = 0$ if $X,Y \in \Gamma E$ we have $K(\partial_k, \partial_\ell) = 0$ $(k, \ell > q)$ so $R^a_{bij} = 0$ unless i or $j < q$. Now

$$\det (I - \Omega^a_b) = 1 + (\text{terms of deg. } 2) + p_1 + (\text{terms of deg. } 6) + + p_2 + \ldots$$

where $p_i \in i^{th}$ Pontrjagin class $\in H^{4i}(Q, Z_2)$. If $2(i_1+\ldots+i_r) > q$ then $p_{i_1} \ldots p_{i_r} = 0$ (since one of $dx^1, \ldots, dx^q$ appears in each nonzero term of Ω^a_b and hence n of $dx^1, \ldots, dx^q$ appear in each term of degree $2n$) so Pont (Q) vanishes in dimensions $> 2q$.

Examples of foliations not obtained from submersions as above are the Reeb foliations of $S^1 \times D^n$, which give also a foliation of S^3 and lead to a foliation of any closed 3-manifold. It is not known if S^5, S^7 etc. have foliations.

There are two results showing the existence of foliations, with hypotheses on the normal bundle much stronger than the conclusion of Bott's theorem:

<u>THEOREM</u> (Wood). M^3 closed, $E \subset TM$ with TM/E trivial and dim. $E = 2 \Longrightarrow E$ homotopic to a foliation.

<u>THEOREM</u> (Phillips). M^n open and TM^n/E^k reduces to a discrete group $\Longrightarrow E^k$ homotopic to a foliation.

<u>REFERENCES</u>:

[1] Bott, R. Some remarks on the obstruction to constructing integrable distributions, (to

appear in Proceedings of AMS Summer Institute on Global Analysis, Berkeley 1968).

(60) Topological equivalence of foliations H. Rosenberg

Let V and V' be manifolds with foliations $\mathfrak{F}$ and $\mathfrak{F}'$ respectively. We say $(V, \mathfrak{F})$ is topologically equivalent to $(V', \mathfrak{F}')$, if there is a homeomorphism $f : V \to V'$ which takes leaves of $\mathfrak{F}$ onto leaves of $\mathfrak{F}'$.

$\mathfrak{F}$ is a Reeb foliation of V^n if V^n is compact, all the leaves of $\mathfrak{F}$ in the interior of V are homeomorphic to R^{n-1}, and if $\partial V^n \neq \phi$, then we demand that ∂V be the union of tori T^{n-1}, each of which is a leaf of $\mathfrak{F}$. Also, we require that $\mathfrak{F}$ be at least of class C^2, and both V and $\mathfrak{F}$ be orientable. A complete topological classification of two dimensional Reeb manifolds (2-manifolds together with a Reeb foliation) has been given by Denjoy: If V^2 is closed and $\mathfrak{F}$ a Reeb foliation of V^2 then V^2 is homeomorphic to T^2 and $\mathfrak{F}$ is topologically equivalent to a linear irrational flow. If $\partial V^2 \neq \phi$ then V^2 is homeomorphic to $S^1 \times [0,1]$ and $\mathfrak{F}$ is topologically equivalent to one of the two foliations sketched in figures 1 and 2:

Figure 1

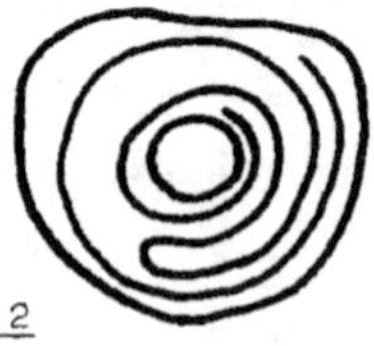

Figure 2

For 3-dimensional manifolds, the following is known:

THEOREM 1. Let $\mathfrak{F}$ be a Reeb foliation of V^3, then $(V^3, \mathfrak{F})$ is topologically equivalent to one of the following:

a) $V = T^3$, and $\mathfrak{F}$ is a linear planar foliation of T^3. More precisely, let $X = (1, a, 0)$ and $Y = (0, 1, b)$, with a and b real numbers, linearly independent over Q. The foliation of R^3 by hyperplanes parallel to the plane spanned by X and Y, passes to a foliation of $T^3 = R^3/Z^3$, and each of the leaves on T^3 is homeomorphic to R^2. This is a linear planar foliation of T^3.

b) $V = S^1 \times D^2$ and $\mathfrak{F}$ is the Reeb foliation of $S^1 \times D^2$.

c) $V = T^2 \times [0,1]$ and $\mathfrak{F}$ is a foliation of V obtained by taking $S^1 \times [0,1]$, together with one of the foliations $\mathfrak{F}_1$ or $\mathfrak{F}_2$, of figures 1 or 2, and defining $\bar{\mathfrak{F}} = \mathfrak{F}_1 \times I$ or $\mathfrak{F}_2 \times I$. Then $\bar{\mathfrak{F}}$ is a foliation of $S^1 \times I \times I$. Now identity $S^1 \times I \times \{0\}$ with $S^1 \times I \times \{1\}$, after making an irrational rotation of $S^1 \times I \times \{1\}$. We can choose $\mathfrak{F}_1$ and $\mathfrak{F}_2$ to be invariant under this rotation, so that $\bar{\mathfrak{F}}$ passes to a Reeb foliation of V. Then $\mathfrak{F}$ is one of the two foliations obtained in this manner.

The proof of these results can be found in [1], [2], [3].

Some related results are:

<u>THEOREM 2</u>. If $n \geq 5$ and V^n is a closed manifold which admits a Reeb foliation, then V^n is homeomorphic to T^n, [3].

<u>THEOREM 3</u>. If V^4 is a closed manifold which admits a Reeb foliation, and if the 3-dimensional Poincaré conjecture holds, then $V^4 = T^4$.

<u>THEOREM 4</u>. If V^3 is a closed simply connected manifold and $\mathfrak{F}$ a foliation of V^3 by planes and tori, then V^3 is homeomorphic to S^3, [2].

<u>THEOREM 5</u>. If V^3 is closed and of rank two (i.e. admits two linearly independent commuting vector fields, but not three), then V^3 is a non trivial 2-torus bundle over S^1 (the converse also holds), [4].

<u>REFERENCES</u>:

[1] Rosenberg, H. Foliations by planes, <u>Topology</u> 7 (1968) 131-138.

[2] Rosenberg, H. and R. Roussarie — Reeb Foliations, Ann. Math. 91 (1970) 1-25.

[3] Rosenberg, H. and R. Roussarie — Topological equivalence of Reeb foliations, (to appear in Topology).

[4] Rosenberg, H., R. Roussarie and D. Weil — A classification of 3-manifolds of rank two, (to appear in Ann. Math.).

(61) Foliations

G. Reeb

Let V, W be compact C^∞ manifolds with $\partial W = \phi$, $\partial V \neq \phi$; let f be a C^∞ function $V \to R$ with $f|_{\partial V} = 0$, $f \neq 0$ elsewhere, and $df \neq 0$ on ∂V, and let ω be a non-vanishing 1-form on W. Then $\alpha = f\omega + df$ defines a non-vanishing 1-form on $V \times W$, and $d\omega \equiv 0 \Rightarrow \alpha \wedge d\alpha = 0$ giving a foliation of $V \times W$ of which $\partial V \times W$ is a leaf. For example $W = S^1$, $\omega = d\theta$, $V = D^n$, $f(x) = |x|^2-1$ gives a well-known foliation of $D^n \times S^1$. If $d\omega \neq 0$ but $\omega \wedge d\omega = 0$ (ω completely integrable) then $d\alpha \wedge d\alpha = 0$; such a ω is known, for example, on S^3, giving α on $D^n \times S^3$ (using f above).

Define a non-vanishing 1-form Ω to be of

class 1 $\Leftrightarrow$ $d\Omega \equiv 0$.

class 2 $\Leftrightarrow$ $\Omega \wedge \Omega \equiv 0$ but $d\Omega \neq 0$ somewhere

class 3 $\Leftrightarrow$ $d\Omega \wedge d\Omega \equiv 0$ but $\Omega \wedge d\Omega \neq 0$ somewhere

class 4 $\Leftrightarrow$ $\Omega \wedge d\Omega \wedge d\Omega \equiv 0$ but $d\Omega \wedge d\Omega \neq 0$ somewhere,

etc.

(Goursat, Cartan, Pfaff).

The existence of a transversely oriented C^∞ foliation on V is equivalent to the existence of a non-vanishing 1-form of class 2 on V.

We write $Cl_\infty V (Cl_\alpha V) = p$ if there is a non-vanishing C^∞ (analytic) 1-form on V of class p (but not of class < p). Given V we ask: what is $Cl_\infty(V)$ or $Cl_\alpha(V)$? We know $Cl_\infty S^3 = 2$, and $Cl_\alpha S^{2n+1}$

≥ 3 (Haefliger). Also S^7 is the union of two copies of $D^4 \times S^3$ on each of which we can construct α as above of class 3, so $Cl_\infty S^7 \leq 3$. Similarly we get $Cl_\infty S^{15} \leq 4$, $Cl_\infty S^{31} \leq 5$, etc., and some results for manifolds of intermediate dimension. The following two theorems are due to Lutz:

<u>THEOREM 1</u> [1]. $Cl_\alpha S^{2n+1} = 3$.

The proof is by direct construction of a suitable 1-form.

<u>THEOREM 2</u>. Let $V_{n,2}$ be the Stiefel manifold of 2-frames in R^n.

(i) Since $V_{n,2}$ fibres over S^{n-1}, if n is even there is a 1-form of class 3 on S^{n-1} which lifts to $V_{n,2}$, so $Cl_\alpha V_{n,2} = 3$.

(ii) If n is odd then $Cl_\infty V_{n,2} \leq 4$ (proved by 'pasting' things together).

From Haefliger we have $Cl_\alpha V_{n,2} \geq 3$, so an outstanding question is: does $Cl_\alpha V_{n,2} = 3$ or 4?

Calabi and Martinet have attempted to mirror the theory of differentiable maps of one manifold into another in the study of manifolds with differential forms. The (local) <u>class</u> of a form plays the role of the <u>rank</u> of a map.

<u>THEOREM 3</u>. (1) [2] If ω is a form with $2 \leq$ degree $\omega \leq n - 2$ $(n \geq 7)$ then ω can be approximated by a form which is of class n everywhere.

(2) If ω is a 1-form it can be approximated by a 1-form with no points where the class $\leq n-\sqrt{2n}$.

<u>REFERENCES</u>:

[1] Lutz, R. Sur la classe maximale des formes de Pfaff sans singularité sur S_{2p+1}, <u>C.R. Acad. Sci. Paris</u> 264 (1967) A1137-A1138.

[2] Martinet, J. Sur les singularités de formes différentielles, (to appear in <u>Ann. Inst. Fourier, Grenoble</u>).

(62) Difféomorphismes du tore T^3 F. Laudenbach

La motivation de cette étude est la suivante:
Si une variété V^4 fermée est feuilletée par des plans R^3, on sait d'après Rosenberg [1] qu'elle est revétue par R^4 et que $\pi_1(V)$ est abélien libre; d'autre part le feuilletage étant défini par une forme fermée sans singularité (Sacksteder [2]), d'après Tischler [4] V est fibrée sur le cercle S^1. Si la conjecture de Poincaré est vraie en dimension 3, la fibre est T^3; alors
$V = T^3 \times [0, 1] / (F(x),1) = (x, 0)$, où F est une difféomorphisme de T^3, induisant l'identité sur le groupe fondamental. Si on montre qu'un tel difféomorphisme est isotope à l'identité, alors V sera difféomorphe à T^4.(*)

Proposition: Tout difféomorphisme $F:T^3 \to T^3$, induisant l'identité sur le groupe fondamental est isotope à l'identité.

(Nous remarquons que ce résultat a aussi été démontré par F. Waldhausen [5]).

Démonstration: On regarde T^3 comme $T^2 \times S^1$, on considère le sous-tore $T = T^2 \times \{0\}$ et on étudie l'intersection de T avec F(T) qui, en position générale, est formée de courbes fermées disjointes, les unes homotopes à zéro et les autres non-homotopes à zéro, mais ces dernières étant homotopes entre elles sur T (resp. sur F(T)) et en nombre pair; si on oriente ces dernières de façon cohérente avec l'homotopie qui les relie, il y en a autant qui ont l'orientation de l'intersection et autant qui ont l'orientation opposée, car l'intersection homologique de T avec F(T) est nulle.

(*) Tout récemment Joubert et Moussu ont démontré que toute variété compacte sans bord de dimension 4, feuilletée par des plans R^3, est difféomorphe à T^4 (à paraître, faculté de Dijon).

En utilisant l'irréductibilité de T^3, on peut éliminer toutes les courbes d'intersections homotopes à zéro. Ensuite on considère une paire (γ, γ'), de deux courbes d'intersection non homotopes à zéro, de signes opposés. Il existe sur F(T) une système de telles courbes limitant une bande A_1 $(= S^1 \times [0, 1])$ ne rencontrant pas T en son intérieur. Dans T, γ et γ' bordent deux bandes A_2 et A'_2; $A_1 \cup A_2$ ou $A_1 \cup A'_2$ borde dans T^3 un tore solide à travers lequel on pourra faire une isotopie de A_1 pour éliminer la paire (γ, γ') ainsi que toutes les paires se trouvant dans A_2 (ou dans A'_2).

Ainsi on peut disjoindre par isotopie F(T) de T. Alors F(T) $\cup$ T sépare T^3 en deux composantes, chacune étant un cobordisme trivial (c'est par exemple une conséquence du théorème de Stallings [3]). On peut donc déformer F jusqu'à ce que F(T) = T; mais la proposition étant vraie pour le tore de dimension 2 (Baer), on se ramène a $F|_T = Id|_T$. En coupant T^3 le long de T, on regarde F comme une difféomorphisme de $T^2 \times [0, 1]$. Par des techniques analogues on peut ramener F à être l'identité sur des sous-variétés convenables le long des quelles on pourra couper; ainsi on arrive à regarder F comme une difféomorphisme du cube égal a l'identité sur le bord. Un tel difféomorphisme est isotope à l'identité relativement au bord (Cerf).

<u>RÉFÉRENCES</u>:

[1] Rosenberg, H. Foliations by planes, *Topology* 7 (1968) 131-138.

[2] Sacksteder, R. Foliations and pseudogroups, *Amer. J. Math.* 87 (1965) 79-102.

[3] Stallings, J. On fibering certain 3-manifolds, *Topology of 3-manifolds and Related Topics (ed. Fort), Prentice Hall, 1961*.

[4] Tischler, D. (To appear).

[5] Waldhausen, F. On irreducible 3-manifolds which are sufficiently large, *Ann. Math.* 87 (1968) 56-88.

(63) Foliations — R. Roussarie

The following 'Schönflies' theorem for 3-manifolds foliated by planes is due to Rosenberg:

THEOREM 1 [1]. Let V be a 3-manifold without boundary, not necessarily compact, with a transversely orientable foliation $\mathfrak{F}$ by planes of class C^k $(k \geq 2)$. Then if $S \subset V$ is a differentiably embedded 2-sphere there exists a differentiably embedded 3-ball D with $\partial D = S$.

From this we can deduce:

THEOREM 2 [2]. If V is a compact 3-dimensional Reeb manifold (see lecture (60)) then V is diffeomorphic to T^3, $D^2 \times S^1$ or $T^2 \times I$.

THEOREM 3 [3]. If V is a compact orientable 3-manifold ($\partial V = \emptyset$) admitting two linearly independent vector fields which commute, then V is a fibre space over S^1 with fibre T^2.

An outline of the proof of Theorem 1 is as follows. First, isotop S into general position with respect to the leaves of $\mathfrak{F}$: then there are a finite number of points where S is not transversal to the leaves, and all such 'critical' points are represented by centres or saddle points on S. The critical points can be arranged to lie on different leaves of $\mathfrak{F}$. We construct on S a vector field X tangent to $S \cap \mathfrak{F}$; using Poincaré-Bendixson theory we prove that the only non-tangent orbits of X are separatrices of saddle points. The proof then consists of a systematic removal by geometric means of all the critical points except two centres.

REFERENCES:

[1] Rosenberg, H. and R. Roussarie — Topological equivalence of Reeb foliations, (to appear in Topology).

[2] Rosenberg, H. and R. Roussarie — Reeb foliations, (to appear in Ann. Math.).

[3] Rosenberg, H., R. Roussarie and D. Weil — A classification of closed 3-manifolds of rank two, (to appear in Ann. Math.).

(64) Algebraic invariants of foliations — B. Reinhart

Let M be a C^∞-manifold of dimension n, with a C^∞ foliation $\mathfrak{F}$ of codimension q. Locally $\mathfrak{F}$ is defined by forms $\omega^1, \ldots, \omega^q$ satisfying $d\omega^\alpha = \sum_{\beta=1}^{q} \theta^\alpha_\beta \wedge \omega^\beta$. Let $\Lambda = C^\infty$ forms on M, and define $\omega\Lambda$ = the ideal $\omega^1\Lambda + \ldots + \omega^q\Lambda$; let $\Omega = \omega^1 \wedge \ldots \wedge \omega^q$. We have $d\Omega = \Theta \wedge \Omega$ where $\Theta = \theta^1_1 + \ldots + \theta^q_q$. There is an exact sequence

$$0 \to \omega\Lambda \overset{i}{\to} \Lambda \overset{R_\Omega}{\longrightarrow} \Lambda\Omega \to 0,$$

where i is inclusion and R_Ω is right multiplication by Ω, giving an exact sequence of chain complexes

$$0 \to (\omega\Lambda, d) \to (\Lambda, d) \to (\Lambda\Omega, d-\Theta) \to 0$$

which induces an exact cohomology sequence

$$\ldots \to H^k(M) \to H^{k+q}(\Lambda\Omega, d-\Theta) \overset{\delta_k}{\to} H^{k+1}(\omega\Lambda, d) \to H^{k+1}(M) \to \ldots .$$

If M is compact, δ_k is Fredholm, and we denote its index by i_k, $k=0,\ldots,n-q$.

In the case q = 1 the sequence reduces to

$$\ldots \to H^k(M) \to H^{k+1}(\Lambda\omega, d-\theta) \overset{\delta_k}{\longrightarrow} H^{k+1}(\omega\Lambda, d) \to H^{k+1}(M) \to \ldots$$

and we obtain an Euler formula $\sum_{\nu=0}^{n-1} (-1)^\nu i_\nu = 0$.

There is a pairing $\Lambda\omega \times \omega\Lambda \to R$ given by $(\alpha \wedge \omega, \omega \wedge \beta) \to \int_M \alpha \wedge \omega \wedge \beta$, and this induces a pairing $H^{n-k}(\Lambda\omega, d-\theta) \times H^{k+1}(\omega\Lambda, d) \to R$. It can be verified that $\ker \delta_{n-k-1} \times \operatorname{im} \delta_k \to 0$, and so there is an induced pairing $\ker \delta_{n-k-1} \times \operatorname{coker} \delta_k \to R$.

Lemma. If $\omega \wedge \beta$ is closed and $\int_M \alpha \wedge \omega \wedge \beta = 0$ for all closed α, then $\{\omega \wedge \beta\} \in \operatorname{im} \delta_k$ ($\dim \alpha = n-k-1, \dim \beta = k$).

Thus $\operatorname{coker} \delta_k$ can be considered as a subspace of the dual space of $\ker \delta_{n-k-1}$, and so $i_k + i_{n-k-1} \geq 0$.

If $\mathfrak{F}$ is given by a fibering $F \to M \to S^1$, the fact that $H^{k+1}(M,F) \cong H^k(F)$ implies that $i_k = 0$ for all k. However, if e.g. $M = S^3$ we get $i_0 = 1$, $i_1 = 0$, $i_2 = -1$, depending only on the homology of S^3. If there were a computation involving the actual foliation we could possibly obtain some non-existence results for foliations of S^5, S^7 etc.. On a torus T^2, $i_k = 0$ for linear foliations. In any case $i_0 = i_1 \geq 0$, so from homology $i_k = 0$ or 1. We can construct a foliation of T^2 with $i_k = 1$.

[As an appendix, a proof was given of Tischler's theorem: If M is compact without boundary, ω is a closed form, and $\varepsilon > 0$, then there is a smooth map $f:M \to S^1$ such that $f^*(dt)$ approximates some integral multiple of ω uniformly within ε. (Here S^1 is parametrized by the real number t modulo 1). In particular if ω is nowhere zero and ε is small enough, then f is a fiber map. (See: D. Tischler, Fibering certain foliated manifolds over S^1, (to appear in Topology).)]

(65) Work of Gromov: generalization of the Smale-Hirsch Theorem — A. Phillips

Let M^n, W^p be C^∞ manifolds, and let Hom (M,W) = C^∞ maps $M \to W$ with the coarse C^1-topology. Let Hom (TM, TW) = fibrewise linear maps $TM \to TW$ with the compact-open topology, and $\operatorname{Hom}^{(n)}(TM, TW)$ = those which are of rank n on each fibre. Define $\operatorname{Imm}(M,W) = d^{-1}\operatorname{Hom}^{(n)}(TM,TW) \subset \operatorname{Hom}(M,W)$, where $d:\operatorname{Hom}(M,W) \to \operatorname{Hom}(TM,TW)$ is the continuous map given by $f \mapsto Tf$.

THEOREM 1 (Smale-Hirsch). $p > n$ ($p \geq n$ if M open) $\Longrightarrow$ $\operatorname{Imm}(M,W) \to$

$\mathrm{Hom}^{(n)}(TM,TW)$ is a weak homotopy equivalence.

Let $\mathcal{O}_M$ be the category of open subsets of M with inclusion maps, $\mathcal{C}$ any category. A functor A: $\mathcal{O}_M \to \mathcal{C}$ is <u>locally defined</u> if $U = U_1 \cup U_2$ (open sets) $\Rightarrow$ ($f \in A(U) \Leftrightarrow i_j^* f \in A(U_j)$, $j=1,2$, i_j=inclusion). A functor A: $\mathcal{O}_M \to$ <u>Top</u> is <u>supple</u> if when $V_2 = V_1 \cup$ (handle) then $A(V_2) \to A(V_1)$ (restriction) has the covering homotopy property (Gromov); we say A is <u>k-supple</u> if the above holds when the index of the handle $\leq k$.

If A is defined on $\mathcal{O}_M$ it can be extended naturally to $\mathcal{S}_M$, the category of subsets of M and inclusion maps.

<u>THEOREM 2</u> (Smale-Thom-Hirsch-Palais-Haefliger-Poenaru theorem - proving machine). Suppose A, B : $\mathcal{S}_M \to$ <u>Top</u> are n-supple and locally defined, and $\Phi : A \to B$ is a natural transformation such that $\Phi : A(D^n) \to B(D^n)$ is a weak homotopy equivalence for $D^n \subset M$ an embedded disk. Then $\Phi : A(M) \to B(M)$ is a w.h.e.. (If M open, (n-1)-supple will do).

<u>Lemma</u> (Gromov). Let A : $\mathcal{S}_M \to$ <u>Top</u> be a subfunctor of Hom (,W) such that (1) A is locally defined, (2) $A(S) \subset \mathrm{Hom}(S,W)$ is open, (3) $\phi \in \mathrm{Diff}(U)$, $f \in A(U) \Rightarrow f\circ\phi \in A(U)$. Then A is (n-1)-supple.

<u>Corollary</u>: If M is open, the following maps are all weak homotopy equivalences:

(i) $\mathrm{Imm}(M,W) = d^{-1}\mathrm{Hom}^{(n)}(TM,TW) \to \mathrm{Hom}^{(n)}(TM,TW)$, $n \leq p$

(ii) $\mathrm{Sub}(M,W) = d^{-1}\mathrm{Hom}^{(p)}(TM,TW) \to \mathrm{Hom}^{(p)}(TM,TW)$, $n \geq p$

(iii) $k\text{-mer}(M,W) = d^{-1}\mathrm{Hom}^{(\geq k)}(TM,TW) \to \mathrm{Hom}^{(\geq k)}(TM,TW)$, $k \leq n,p$.

(The functors Imm(,W), $\mathrm{Hom}^{(n)}(T\ ,TW)$ etc. all satisfy the hypotheses of Gromov's lemma.)

Note that (i) is a special case of Theorem 1. In fact (ii) implies the stronger Theorem 1. Also (3) holds for M compact ($k < p$) but this does not seem to follow from Gromov's lemma.

<u>Applications to foliations</u>: Suppose W has a distribution σ of k-planes, and let $Hom^{\sigma}(TM,TW)$ = fibrewise linear maps transverse to σ. If $Trans^{\sigma}(M,W) = d^{-1}Hom^{\sigma}(TM,TW)$ the functor $Trans^{\sigma}(\ ,W)$ satisfies Gromov's lemma. We deduce

<u>THEOREM 3</u>. If M is open, $Trans^{\sigma}(M,W) \to Hom^{\sigma}(TM,TW)$ is a w.h.e.. This holds in particular in case W carries a foliation $\mathfrak{F}$ and $\sigma = T\mathfrak{F}$. Note that $g \in Trans^{T\mathfrak{F}}(M,W)$ pulls back $\mathfrak{F}$ to a foliation $g^*\mathfrak{F}$ of M. One obtains

<u>Corollary</u>. M open, $\sigma^{\perp}$ has a discrete structural group $\Rightarrow$ σ is homotopic to an integrable distribution.

Gromov has also considered functors whose definition depends on higher partial derivatives, and has obtained interesting results of a differential-geometric nature.

LIST OF SPEAKERS AT THE AFTERNOON SESSIONS

OF THE SUMMER SCHOOL 15th - 25th JULY 1969

SUMMER SCHOOL LECTURES

(1) Foliations of codimension one G. Reeb

A few properties of foliations of codimension one were discussed. The main topic was the property of complete stability which may be stated thus:

THEOREM. Let V^n be a (C^∞) compact manifold, and ω be a Pfaffian form with $\omega \neq 0$ at each point of V^n and $\omega \wedge d\omega \equiv 0$; let V^{n-1} be a compact, simply connected leaf of the equation $\omega = 0$. Under these assumptions all the leaves of $\omega = 0$ are the fibres of a fibre bundle over S^1 and V^{n-1} is the fibre.

The proof of the theorem is to be found in [1].

The situation described in the theorem has a very strong kind of (C^0) structural stability: if we replace the form ω by ω' which is C^0 close to ω, and if ω' satisfies the hypotheses of the theorem (i.e. if $\omega' \neq 0$ at each point and $\omega' \wedge d\omega \equiv 0$) then the equation $\omega' = 0$ defines a fibration of V^n isomorphic to the fibration defined by $\omega = 0$.

A consequence of the theorem is the following.

Corollary. Let $\alpha \equiv \Sigma a_i(x)dx^i$ be a complex analytic Pfaffian form in some neighbourhood of the origin of $\mathbb{C}^n$, with the following assumptions:

a) $n \geq 3$, $a_i(0) = 0$, $\left(\frac{\partial a_i}{\partial x^j}\right)_{x=0}$ is a matrix of rank n,

b) $\alpha \wedge d\alpha \equiv 0$.

Then there exists a holomorphic (non-constant) function f(x) in a neighbourhood of the origin such that $df \wedge \alpha \equiv 0$ (i.e. f is a first integral of $\alpha = 0$).

A proof of the corollary is also to be found in [1]. Painlevé [2] asked the following question: how can the results of

Painlevé concerning equation $f(y,y',x) = 0$ be extended to Pfaffian (and completely integrable) equations $\omega = 0$ in the complex domain? The corollary shows that, under some mild restriction, the fixed singularities of ω do not give rise to non-algebraic singularities of the solutions, as does happen in the case of $f(y,y',x) = 0$.

This leads also to the following conclusion: the work of Painlevé on ordinary differential equations in the complex domain (and the suggested generalisation by Painlevé) gives an a posteriori justification for the investigation of foliations.

REFERENCES:

[1] Painlevé, P. Leçons de Stockholm, Paris, 1904.

[2] Wu wen-tsün and G. Reeb Sur quelques propriétés topologiques des espaces fibrés et des variétés feuilletées, Hermann et Cie., Paris, 1952.

(2) Expanding attractors R. Williams

Suppose $f:M \to M$ is a diffeomorphism and Λ is an attractor for f which has a hyperbolic structure, $E^u + E^s$. (Technical terms are as described in [3]; see also lecture notes by E.C. Zeeman (to appear).) We use u and s also to denote the dimensions of the fibres of the bundles E^u, E^s, and explicate the special case:

Definition. Λ is an expanding attractor provided $\dim \Lambda = u$.

We define branched n-manifold, and a self immersion $g:K \to K$ of a branched n-manifold K is assumed to satisfy three axioms: (1) the non-wandering set of g is all of K; (2) each $x \in K$ has a neighbourhood N such that $g^i(N)$ lies in an n-disk for some i; and (3) g is an expanding map. Such a (g,K) is a presentation of an n-solenoid Σ and shift map $h : \Sigma \to \Sigma$, where Σ is the inverse limit of the sequence

$$K \xleftarrow{g} K \xleftarrow{g} K \xleftarrow{g} \cdots$$

and for $x = (x_0, x_1, x_2, \ldots) \in \Sigma$, $h(x_0, x_1, \ldots) = (gx_0, gx_1, \ldots)$.

Strong use is made of the (generalized) stable manifolds $W^s(x,f)$, $x \in \Lambda$, as formulated by Smale in [3] and shown to exist by Hirsch and Pugh in [1].

STATEMENT OF RESULTS

THEOREM A. Assume Λ is an expanding attractor for $f \in \mathrm{Diff}(M)$ and that the foliation $\{W^s(x,f) : x \in \Lambda\}$ is C^1 on some neighbourhood of Λ. Then $f|\Lambda$ is conjugate to the shift map h of a solenoid Σ, i.e. there is a homeomorphism $\phi : \Lambda \to \Sigma$ such that $f|\Lambda = \phi^{-1}h\phi$.

THEOREM B. Assume $h : \Sigma \to \Sigma$ is a shift map of an n-solenoid. Then there is a manifold M and $f \in \mathrm{Diff}^r(M)$ having an expanding attractor Λ such that $f|\Lambda$ is conjugate to h.

THEOREM C. Each point of an n-solenoid has a neighbourhood of the form (Cantor set) × (n-disk).

THEOREM D. The periodic points of a shift map of an n-solenoid are dense.

THEOREM E. The reduced cohomology over Z of an n-solenoid is not 0.

This last is all the author has proved at this time toward a very strong

Conjecture. $H^n(\Sigma;Z) \neq 0$, $H^1(\Sigma;Z) \neq 0$, and $H^1(\Sigma;Z)$ generates $H^*(\Sigma;Z)$ as a ring. If orientable, Σ is a fibre bundle over an n-manifold with Cantor set as fibre.

Complete proofs will appear elsewhere. Meanwhile a detailed outline is available as [3].

REFERENCES:

[1] Hirsch, M. and C. C. Pugh — Stable manifold theorems (to appear in Proceedings of AMS Summer Institute on Global Analysis, Berkeley 1968).

[2] Smale, S. Differentiable dynamical systems, Bull. A.M.S. 73 (1967) 747-817.

[3] Williams, R. Expanding attractors (a detailed outline), (to appear in the Mt. Aiguall Conference on Differential Topology (Spring 1969)). (Memo notes, Northwestern University.)

(3) Equivalence of dynamical systems S. A. Robertson

This seminar was concerned with an elementary discussion of a few of the more obvious equivalence relations that can be defined between flows on topological spaces, manifolds, or linear spaces. Here we list seven such relations and state their mutual relationships, which are established by a combination of simple arguments and pathological counterexamples.

In some cases (e.g. $G \to F$ and $Q \to 0$), implications can be established for flows of special types (e.g. by insisting that the space be a manifold, or that the flow have no rest-points, etc.).

In what follows, ϕ and ψ are flows on topological spaces X and Y.

F. ϕ is flow-equivalent to ψ iff there exist an isomorphism $f : R \to R$ and a homeomorphism $h : X \to Y$ such that $\psi \circ (f \times h) = h \circ \phi$.

0. Put $x \sim x'$ iff there exists $t \in R$ with $\phi(t,x) = x'$. The $\sim$ classes are the orbits of ϕ, and there is a quotient map $\pi_\phi : X \to X_\phi$ onto the orbit-space $X_\phi = X/\sim$. ϕ is orbit-equivalent to ψ iff there exist a homeomorphism $h : X \to Y$ and a map $h' : X_\phi \to X_\psi$ such that $h' \circ \pi_\phi = \pi_\psi \circ h$.

00. A map $k : X \to Y$ preserves (reverses) the orientation of the orbit through $x \in X$ iff there is an increasing (decreasing) homeomorphism $\alpha : R \to R$ such that, for all $t \in R$, $k(\phi(t,x)) = \psi(\alpha(t), k(x))$. ϕ is oriented-orbit-equivalent to ψ iff there is an orbit-equivalence (h,h') such that h preserves the orient-

ation of every orbit of ϕ. (Remark: If $h : X \to Y$ is a homeomorphism, and γ is an orbit of ϕ, then h may preserve or reverse the orientation of γ, may do both (if γ = rest-point) and may even do neither (X must then be non-Hausdorff).)

G. ϕ determines a continuous homomorphism $h_\phi : R \to H(X)$ into the group of homeomorphisms of X (compact-open topology), with $h_\phi(t)(x) = \phi(t,x)$. ϕ is <u>group-equivalent</u> to ψ iff there are topological group isomorphisms $f : R \to R$ and $\xi : H(X) \to H(Y)$ such that $h_\psi \circ f = \xi \circ h_\phi$.

Q. ϕ is <u>quotient-equivalent</u> to ψ iff there is a homeomorphism $\xi : X_\phi \to X_\psi$.

C^rF. If X and Y are C^r-manifolds, ϕ is C^r <u>flow-equivalent</u> to ψ iff there is a flow-equivalence (f,h) from ϕ to ψ for which h is a C^r-diffeomorphism. Here $r \geq 1$.

S. If X and Y are (finite-dimensional) linear spaces, and ϕ and ψ are linear flows generated by the linear endomorphisms ('principal velocities') $V_\phi : X \to X$, $V_\psi : Y \to Y$, then ϕ is <u>similar</u> to ψ iff there is a linear isomorphism $\theta : X \to Y$ such that $\theta \circ V_\phi = V_\psi \circ \theta$.

<u>THEOREM</u>: The above relations are connected by the following implications, where applicable.

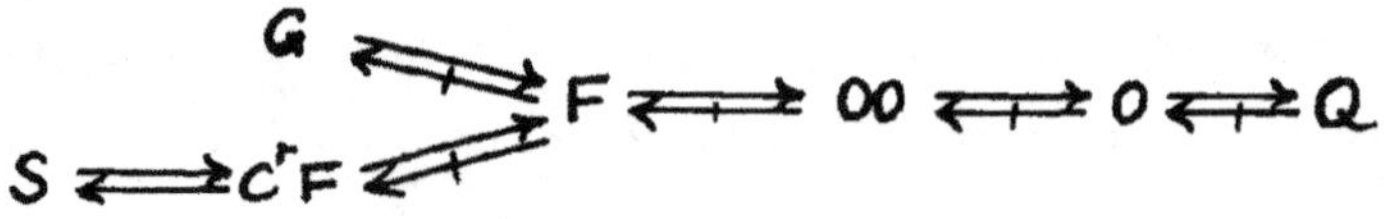

(4) <u>Generic bifurcation</u> <u>K. Meyer</u>

The study of Hamiltonian systems of ordinary differential equations leads naturally to the study of one parameter families of area-preserving diffeomorphisms. The existence of periodic solutions is reduced to the study of periodic points of the diffeomorphisms.

Let $\mathfrak{F}$ be the space of all C^∞ maps $Q : M \times S \to M$ where M is a 2-dimensional symplectic manifold, S is the unit circle and $Q(\cdot,s) : M \to M$ is symplectic. We define a set $\mathcal{G} \subset \mathfrak{F}$ and show that $\mathcal{G}$ is generic (in the sense of Baire category theory) in $\mathfrak{F}$. The set $\mathcal{G}$ is characterized by a precise description of the nature of the fixed points and periodic points of $\phi \in \mathcal{G}$. The characterization of $\phi \in \mathcal{G}$ gives information about 1) the dependence of periodic points of ϕ on the parameter $s \in S$, 2) the nature of the bifurcation of the periodic points and 3) the stability and instability of the periodic points.

(5) Probabilistic convergence of approximations for partial differential equations — H. Kushner

We consider the convergence of numerical methods for the partial differential equation

(1) $\mathcal{L}V = k(x)$, $\qquad V(\partial G) = \phi(\partial g)$

$$\mathcal{L} = \sum_{ij} a_{ij} \frac{\partial^2}{\partial x_i \partial x_j} + \Sigma f_i \frac{\partial}{\partial x_i}, \qquad x = (x_1, \ldots, x_r)$$

where G is a suitable open set with boundary ∂G, k and ϕ are continuous and $\{a_{ij}\}$ is non-negative definite for all x_j, and f_i and σ_{ij} satisfy a Lipschitz condition and are bounded ($a_{ij} = \sum_k \sigma_{ik}\, \sigma_{jk}$). If $\mathcal{L}$ is only semi-definite, or if G has corners, then classical numerical analysis, which requires that V have at least 2 continuous derivatives, yields little information on convergence of finite difference approximations to (1), as the finite difference interval $h \to 0$. In fact V may be only a weak solution.

For further clarification consider the (Itô) stochastic differential equation

(2) $dx = f(x)dt + \sigma(x)dz$, z_t a vector Wiener process, and let $\tau = \inf\{t: x_t \notin G\}$. Then, if

(3) $W(x) \equiv E_x \int_0^{\tau} k(x_s)ds + E_x \phi(x_\tau)$

has continuous second derivatives, it satisfies (1) (with $W(x) = \lim_{y \to x} E_y \phi(x_\tau)$ for $x \in \partial G$, $y \in G$). Conversely, if a smooth solution to (1) exists, it has the representation (3). Thus, whether or not a smooth solution to (1) exists, we consider (3) as its solution (it is a weak solution), and investigate whether $V_h(x)$, the finite difference solution, converges to $W(x)$ as $h \to 0$.

Next the particular finite difference approximations are used: Define the grid $R_h = \{(n_1 h, n_2 h, \ldots, n_r h), n_i = 0, \pm 1, \ldots\}$, and $G_h = R_h \cap G$. For convenience of exposition <u>only</u>, let $a_{ij} = 0$ for $i \neq j$. Let e_i = unit vector in i^{th} direction and

$$V_{x_i x_i}(x) \to [V(x + e_i h) - 2V(x) + V(x - e_i h)]/h^2$$

$$V_{x_i}(x) \to \frac{1}{h} \begin{Bmatrix} [V(x + e_i h) - V(x)] \\ [V(x) - V(x - e_i h)] \end{Bmatrix} \quad \begin{matrix} \text{if } f_i \geq 0 \\ \text{if } f_i < 0. \end{matrix}$$

Denoting the solution to the f.d. equation by $V_h(x)$, and collecting terms, gives (where upper entry is used if $f_i \geq 0$, and lower if $f_i < 0$) for $x \in G_h$,

(5) $$V_h(x) = \sum_i \frac{V_h(x+e_i h)}{Q_h(x)} \begin{Bmatrix} h|f_i| + a_{ii} \\ a_{ii} \end{Bmatrix} + \sum_i \frac{V_h(x-e_i h)}{Q_h(x)} \begin{Bmatrix} a_{ii} \\ h|f_i| + a_{ii} \end{Bmatrix} + \frac{h^2 k(x)}{Q_h(x)},$$

$$Q_h(x) = 2\sum_i a_{ii} + \sum_i h|f_i|$$

and $V_h(x) = \phi(x)$ for $x \notin G_h$. Since the coefficients are non-negative and sum to at most unity (and can be defined for all $x \in R_h$) we rewrite (5) as (for $x \in G_h$)

(6) $V_h(x) = \sum_i V_h(x+e_ih)p_h(x,x+e_ih) + \sum_i V_h(x-e_ih)p_h(x,x-e_ih)+\rho_h(x)k(x)$,

where the $p_h(x,y)$ are transition probabilities for a Markov chain $\{\xi_n^h\}$ on R_h.

Assume (A1) $P_x\{T_h < T\} \geq \varepsilon > 0$

where $T_h = \inf\{n: \xi_n^h \notin G_h\}$ and T and ε are some real numbers. ((A1) is typical in examples and occurs if $\sigma'\sigma = \Sigma$ has a submatrix with full rank for all x). Then (6) has the solution

(7) $V_h(x) = E_x \sum_{n=0}^{T_h-1} k(\xi_n^h)\ \rho_h(\xi_n^h) + E_x\phi(\xi_{T_n}^h)$.

Using some limit theorems for sequences of measures corresponding to Markov processes, it is shown that $V_h(x) \to W(x)$ under the additional hypothesis

(A2) all points on ∂G are either regular (in sense of Dynkin) or are inaccessible.

(A2) is also typical in examples.

(6) Mathematical structure of network synthesis — R. Brockett

Let $\Phi_A(t,\sigma)$ denote the solution of

$$\frac{d}{dt}\,\Phi(t,\sigma) = A(t)\Phi(t,\sigma)\ ;\ \Phi(\sigma,\sigma) = I$$

where I is the identity matrix. Let prime denote matrix transpose and let $\Sigma(m,n)$ be a square matrix

$$\Sigma(m,n) = \begin{pmatrix} I_m & 0 \\ 0 & -I_n \end{pmatrix}.$$

External descriptions of linear electrical networks take the form of an integral equation

$$y(t) = W_0(t)\ u(t) + \int_{t_0}^{t} W(t,\sigma)\ u(\sigma)\ d\sigma + y_0(t)$$

where u and y are m-vectors having the decomposition

$$u = \begin{bmatrix} v_1 \\ i_4 \end{bmatrix} \quad ; \quad y = \begin{bmatrix} i_1 \\ v_4 \end{bmatrix}$$

where v_1 and i_1 are α-dimensional vectors of voltages and currents at a set of α terminal pairs and v_4 and i_4 are δ-dimensional vectors of voltages and currents at a second set of δ terminal pairs. W_0 and W are smooth and, because of the character of networks, we have

$$\Sigma(\alpha,\delta)\ W_0\ \Sigma(\alpha,\delta) = W_0' \quad ; \quad \Sigma(\alpha,\delta)\ W\ \Sigma(\alpha,\delta) = W'.$$

Internal descriptions of networks take the form of an ordinary differential equation and an equation for y

(1) $\quad \dot{x} = Ax + Bu \quad ; \quad y = Cx + Du.$

If the energy storage elements are time invariant then these equations can, with no loss of generality, be assumed to have the form

$$(2) \quad \begin{bmatrix} \Sigma(\beta,\gamma) & 0 \\ 0 & \Sigma(\alpha,\delta) \end{bmatrix} \begin{bmatrix} A & B \\ C & D \end{bmatrix} \begin{bmatrix} \Sigma(\beta,\gamma) & 0 \\ 0 & \Sigma(\alpha,\delta) \end{bmatrix} = \begin{bmatrix} A & B \\ C & D \end{bmatrix}' .$$

A basic question in network synthesis concerns the characterization of all internal descriptions which give rise to the same external description. Clearly the relationship between the internal and external description is

$$W_0 = D \ ; \quad W(t,\sigma) = C(t)\ \Phi(t,\sigma)\ B(\sigma).$$

One procedure for generating new realizations from a given one is to make a linear transformation on x, say x = Tx. Then

(3) $\quad \dot{z} = (TAT^{-1} + \dot{T}T^{-1})\ z + TBu \ ; \quad y = CT^{-1}z + Du.$

Of course this corresponds to a network only in the case where

$$(4) \quad \begin{bmatrix} \Sigma(\beta,\gamma) & 0 \\ 0 & \Sigma(\alpha,\delta) \end{bmatrix} \begin{bmatrix} TAT^{-1}+\dot{T}T^{-1} & TB \\ CT^{-1} & D \end{bmatrix} \begin{bmatrix} \Sigma(\beta,\gamma) & 0 \\ 0 & \Sigma(\alpha,\delta) \end{bmatrix} = \begin{bmatrix} TAT^{-1}+\dot{T}T^{-1} & TB \\ CT^{-1} & D \end{bmatrix}' .$$

We say an internal description is minimal if there exists no other internal description having a lower dimensional differential equation and the same external description. Direct characterizations of minimality are well known. The question we want to focus on is that of characterizing what T's satisfy the symmetry constraints of equation (3). A partial answer is given by the following theorem. (Compare with work of Youla and Tissi cited in [1]).

THEOREM 1. If T is constant and if (1) is a minimal internal description satisfying the symmetry conditions (2) then the internal description (3) will satisfy the symmetry conditions (4) if and only if $T'\Sigma(\beta,\gamma)T = \Sigma(\beta,\gamma)$.

Clearly the set of solutions of $T'\Sigma T = \Sigma$ forms a group (a pseudo-orthogonal group preserving the pseudo-length $x_1^2 + x_2^2 + \dots x_\beta^2 - x_{\beta+1}^2 - x_{\beta+2}^2 \dots - x_{\beta+\alpha}^2$). In some cases this theorem can be interpreted as implying the difference between the electric field energy and the magnetic field energy is the same for all realizations.

REFERENCES:

[1] Brockett, R.W. and R.A. Skoog — A new perturbation theory for the synthesis of nonlinear networks, (to appear in Proceedings of the American Mathematical Symposium on Electrical Network Theory, April 1969).

(7) Actions of R^2 on manifolds — H. Rosenberg

We consider C^∞ actions $\phi: R^2 \times V \to V$ of R^2 on a manifold V. Let $\mathfrak{a}_2(V)$ denote the set of all such actions. The following was proved by Lima in 1962.

THEOREM 1. If V is 2-dimensional, compact and $\chi(V) \neq 0$ then any

$\phi \in \mathcal{A}_2(V)$ has a fixed point.

Problem I. Does $\chi(V) \neq 0 \Rightarrow$ any $\phi \in \mathcal{A}_2(V)$ has a fixed point?

It is a basic problem to try to approximate $\phi \in \mathcal{A}_2(V)$ by actions with 'nice' singularities - and to define what is meant by 'nice'.

Problem II. Suppose $\phi \in \mathcal{A}_2(V)$ has no fixed point. Does there exist $\phi' \in \mathcal{A}_2(V)$ having no fixed point and with a finite (or countable) number of 1-dimensional orbits? $\{II \Rightarrow I\}$.

Define ϕ to be non-singular $\Leftrightarrow$ every orbit has dimension 2. Then ϕ non-singular $\Rightarrow$ orbits of ϕ give a foliation of V whose leaves are R^2, $S^1 \times R$ or T^2.

If there exists a non-trivial action $\phi \in \mathcal{A}_2(V)$ then differentiation gives two non-trivial commuting vector fields on V. The converse is false, e.g. $V = R^3 - \{0\}$ has three commuting non-zero vector fields but no non-trivial $\phi \in \mathcal{A}_2(V)$.

Define a 3-manifold V^3 to be irreducible if every $S^2 \subset V^3$ bounds a D^3.

THEOREM 2. If there exists a non-singular action on V^3 (compact or not) then V^3 is irreducible.

This is a generalization of

THEOREM 3. If V^3 admits a transversely oriented C^2 foliation $\mathcal{F}$ whose leaves are R^2, then V^3 is irreducible.

Outline of proof (see p.118, these Proceedings): Given $S^2 \subset V^3$ we first put it into general position with respect to $\mathcal{F}$, and then observe the foliation $g = \{L \cap S^2 \mid L \in \mathcal{F}\}$ on S^2. We can assume the singularities of g lie on different leaves of $\mathcal{F}$. Using Poincaré-Bendixson theory on S^2 and the idea of holonomy it is easy to prove that all leaves of g are closed. A detailed geometrical argument then shows how to remove all singularities except two, and the theorem of Reeb that a disc on a leaf of $\mathfrak{F}$ can be 'lifted' to neighbouring leaves is used to finish the proof.

The proof of Theorem 2 is similar, although additional arguments due to Reinhart and Lima are needed to prove that the leaves of g are closed when the leaves of $\mathfrak{F}$ may be R^2, $S^1 \times R$ or T^2.

If V^3 is closed and orientable, then the existence of two commuting non-trivial vector fields is in fact equivalent to the existence of non-trivial $\phi \in \mathcal{O}_2(V^3)$.

The following result is due to Rosenberg, Roussarie and D. Weil.

THEOREM 4. If $\phi \in \mathcal{O}_2(V)$ is non-singular then V is a T^2-bundle over S^1.

(8) A generalization of Mackey's imprimitivity theorem — K. Parthasarathy

The equation of motion of a free particle in relativistic quantum mechanics is derived as follows: Consider the restriction of an irreducible projective unitary representation of the inhomogeneous Lorentz group or the Poincaré group to the one parameter subgroup of time translations. This yields a one parameter group of unitary operators whose infinitesimal generator yields the equation of motion. The Dirac equation for the electron can be derived in this manner (cf. [1]).

An application of Wigner's unitarity antiunitarity theorem (see [4]) suggests that representations of both the unitary and antiunitary operators should be considered in the above context. There does not seem to exist such a general theory of unitary antiunitary representations. The main purpose of this lecture is to announce a generalisation of Mackey's imprimitivity theorem [2] in this context.*

Let G be a locally compact second countable group, $H \subset G$ a closed subgroup and $X = G/H$ the homogeneous space of right cosets. An imprimitivity system for G on X is a triple $\{\mathcal{H}, U_g, P(E)\}$ where $\mathcal{H}$ is a complex separable Hilbert space, $g \to U_g$ is a projective unitary antiunitary representation of G, $E \to P(E)$ is a projection valued measure on the Borel field $\mathcal{B}$ of X and

$$U_g P(E) U_g^{-1} = P(gE) \quad \text{for all } g \in G,\ E \in \mathcal{B} .$$

U_g satisfies the equation

$$U_{g_1} U_{g_2} = \sigma(g_1, g_2)\, U_{g_1 g_2} .$$

σ is called the multiplier of the representation $g \to U_g$. The imprimitivity system is said to be irreducible if there is no proper closed subspace invariant under all U_g and $P(E)$.

Let $x_0 \in X$ be the point corresponding to the coset H and γ be a one-one Borel map from X into G such that $\gamma(x)x_0 = x$ for all $x \in X$. Let μ be a quasi invariant measure on X, i.e. $\mu(E) = 0$ if and only if $\mu(gE) = 0$ for all $g \in G$. For every finite or countable cardinal n, let $L_2(\mu,n)$ be the direct sum of n copies of $L_2(\mu)$. Any element of $L_2(\mu,n)$ can be written as $\underset{\sim}{f}(x) = (f_1(x), f_2(x), \ldots)$ where for each fixed x, f(x) can be considered as an element of the complex Hilbert space $\mathbb{C}^n$ or ℓ_2 according as n is finite or infinite. Let $E \to P^0(E)$ be the canonical projection valued measure in $L_2(\mu,n)$ defined by $P^0(E)f = (\chi_E f_1, \chi_E f_2, \ldots)$ where χ_E is the characteristic function of the set E.

---

* A proof of this may be found in [3].

For a given imprimitivity system let $G^+ = \{g: U_g \text{ is unitary}\}$ and $G^- = \{g: U_g \text{ is antiunitary}\}$.

THEOREM 1. Let $\{\mathcal{H}, U_g, P(E)\}$ be an imprimitivity system for G on X with multiplier σ. Suppose G^+ acts transitively on X. Then there exists an unitarily equivalent system $\{L_2(\mu,n), V_g, P^0(E)\}$ where

$$(V_g f)(x) = \frac{\sigma(g,\gamma(g^{-1}x))}{\sigma(\gamma(x),\gamma(x)^{-1}g\gamma(g^{-1}x))}\left\{\left(\frac{d\mu}{d\mu^g}\right)(g^{-1}x)\right\}^{\frac{1}{2}} M_{\gamma(x)^{-1}g\gamma(g^{-1}x)} f(g^{-1}x)$$

where $h \to M_h$ is an n dimensional projective unitary antiunitary representation of H (in $\mathbb{C}^n$ or ℓ_2 according as $n < \infty$ or $n = \infty$) with multiplier σ, and μ^g is the measure defined by $\mu^g(E) = \mu(gE)$. The imprimitivity system is irreducible if and only if the representation $h \to M_h$ is irreducible.

Now suppose G^+ does not act transitively. Then the G^+ action has two orbits X^+ and X^- in X. Further $H \subset G^+$. Let μ^+ be the restriction of μ to X^+. For any $E \subset X$, let $E^+ = E \cap X^+$, $E^- = E \cap X^-$. Let $g_0 \in G^-$ be chosen and fixed. Consider the Hilbert space $L_2(\mu^+,n) \oplus L_2(\mu^+,n)$. Any operator in this can be denoted by a second order matrix whose entries are operators in $L_2(\mu^+,n)$. Let $E \to P_0(E)$ be the projection valued measure defined by

$$P_0(E) = \begin{pmatrix} P^0(E^+) & 0 \\ 0 & P^0(g_0E^-) \end{pmatrix}.$$

THEOREM 2. Let $\{\mathcal{H}, U_g, P(E)\}$ be an imprimitivity system for G on X with multiplier σ. Suppose G^+ does not act transitively on X. Then there exists a unitarily equivalent system $\{L_2(\mu^+,n) \oplus L_2(\mu^+,n), V_g, P_0(E)\}$ where

$$V_g = \begin{pmatrix} V_g^+ & 0 \\ 0 & \overline{\alpha(g)}\, s^+\, V^+_{g_0 g g_0^{-1}}\, s^+ \end{pmatrix} \text{ if } g \in G^+,$$

$$V_{g_0} = \begin{pmatrix} 0 & s^+ \\ \overline{\sigma(g_0,g_0)}\, s^+\, V^+_{g_0^2} & 0 \end{pmatrix},$$

$$\alpha(g) = \sigma(g_0,g)\,\overline{\sigma(g_0,g_0^{-1})}\,\sigma(g_0 g,g_0^{-1}),\ g \in G^+,$$

$g \to V^+_g$ is an induced representation of G^+ induced by a projective unitary representation $h \to L_h$ of H with multiplier σ and s^+ is the complex conjugation in $L_2(\mu^+,n)$ which sends an element f to $\bar{f}$. Since G^+ and g_0 generate G, V_g is completely determined by the above formulae. The imprimitivity system described above is irreducible if and only if the representation L of H is irreducible.

Remark. Theorems 1 and 2 constitute a complete generalization of Mackey's celebrated imprimitivity theorem when antiunitary operators are present. This also leads to a classification of all the projective unitary antiunitary representations of the Poincaré group and thus solves some of the questions raised in [1].

REFERENCES:

[1] Foldy, L. Synthesis of covariant particle equations, Phys. Rev. 103 (1956) 568-581.

[2] Mackey, G.W. Unitary representations of group extensions I, Acta Math. 99 (1958) 265-311.

[3] Parthasarathy, K. Projective unitary antiunitary representations of locally compact groups, Comm. Math. Phys. 15 (1969) 305-328.

[4] Varadarajan, V.S. Geometry of Quantum Theory, Vol. I, Van Nostrand, Princeton, 1968.

(9) Algebraic problems in dynamical systems — R.E. Kalman

The purpose of this lecture is to explicate some of the algebraic machinery used to establish the isomorphism between the external description of a dynamical system (= "black box") and the corresponding internal description (= "classical" dynamical system). In particular, it is of interest to relate the structure of the (internal) state space to that of the (external) input/output map. The constructions may be summarized in the following form:

THEOREM. The state set X_f of a linear [multilinear] input/output map f over an arbitrary field k admits the following structure:

(i) X_f is a k-vector space [a k-affine variety];

(ii) X_f is a k[z]-module.

REFERENCES:

[1] Kalman, R.E., P.L. Falb and M.A. Arbib — Topics in mathematical system theory, McGraw-Hill, 1969.

[2] Kalman, R.E. — Lecture notes for C.I.M.E. Summer School on Controllability, Bologna (Italy) 1968. (Edizioni Cremonese, 1969).

[3] Kalman, R.E. — Lectures on algebraic system theory, Springer Lecture Notes in Mathematics, 1970.

(10) Almost periodic minimal sets — Dame Mary Cartwright

Consider the autonomous system $\dot{x} = \psi(x)$ $(x \in R^n)$, where ψ satisfies conditions sufficient to ensure that a solution $x = \phi(t,x_0)$, $t \in (-\infty,\infty)$, $x_0 = \phi(0,x_0)$ exists and is uniquely determined by x_0 and continuous in x_0 for all values considered. Define ϕ to be recurrent if for every $\varepsilon > 0$, $T > 0$ there exists $t(\varepsilon) > T$ such that $||x_0 - \phi(t(\varepsilon),x_0)|| < \varepsilon$. It is well known that for $n = 2$ the only recurrent solutions are periodic, but for $n = 3$ Denjoy has shown that on a torus there are three types of recurrent solution: (a) periodic solutions, (b) quasi-periodic solutions $x = \Phi(\lambda_1 t,\lambda_2 t)$

(λ_1/λ_2 irrational, Φ continuous, of period 2π in both variables), (c) solutions not almost periodic. If ψ is sufficiently smooth, case (c) can be excluded on a 2-torus, but no smoothness hypothesis can exclude non-a.p. solutions in general if $n \geq 4$. We therefore study almost periodic solutions.

A vector function $\chi(t)$ is <u>uniformly almost periodic</u> if

$$A(\lambda) = \lim_{T\to\infty} \frac{1}{T}\int_0^T \chi(t)e^{-\lambda t}\, dt = 0$$

except for an enumerable set $\lambda = \Lambda_1, \Lambda_2, \ldots$; we write

$\chi(t) \sim \sum_{\nu=1}^{\infty} A_\nu e^{i\Lambda_\nu t}$, $A_\nu = A(\Lambda_\nu) \neq 0$. Let λ_j $(j = 1,2,\ldots) \in R$ be such that if r_j is rational then

$$\sum_{j=1}^{J} r_j\lambda_j = 0 \Rightarrow r_1 = r_2 = \ldots = r_J = 0.$$

We can express $\Lambda_\nu = \sum_{j=1}^{J(\nu)} r_j^{(\nu)}\lambda_j$, $\nu = 1,2,\ldots$, in which case the λ_j are <u>basic frequencies</u> and $\{\lambda_j\}$ is a <u>base</u>. These are not unique, but we define a standard base by $\lambda_1 = \Lambda_1$, $\nu_1 = 1$, ν_J = greatest integer such that $\Lambda_\nu = \sum_{j=1}^{J-1} r_j^{(\nu)} \lambda_j$, $\nu = 1,2,\ldots,\nu_j-1$, $r_j^{(\nu)}$ rational, and put $\Lambda_{\nu_J} = \lambda_J$, $J = 2,3,\ldots$.

The series for χ may not converge, but there is a spatial extension of $\chi(t)$ defined by $\Phi(\tau_1,\tau_2,\ldots) = \lim_{k\to\infty} \Phi_k(\tau_1,\tau_2,\ldots,\tau_{J(k)})$

where $\quad \Phi_k(\tau_1,\tau_2,\ldots,\tau_{J(k)}) = \sum_{\nu=1}^{N(k)} b_j^{(k)} \exp\left(i \sum_{j=1}^{J(\nu)} r_j^{(\nu)} \lambda_j\tau_j\right).$

The limit is uniform for $\tau_j \in (-\infty,\infty)$, and $b_\nu^{(k)} \to A_\nu$ as $k \to \infty$ for each ν. Obviously $\chi(t) = \Phi(t,t,\ldots)$, and it is well known that the values of $\chi(t)$ are dense in $\{\Phi(\tau_1,\tau_2,\ldots) | \tau_j\in(-\infty,\infty),\ j=1,2,\ldots\}$. Note that if $r_j^{(\nu)} = p_j^{(\nu)}/q_j^{(\nu)}$ (where $p_j^{(\nu)}$, $q_j^{(\nu)}$ coprime) $q_j^{(\nu)}$ may

$\to \infty$, but if $q_j(k)$ is the LCM of $q_j^{(\nu)}$ ($\nu = 1,2,\ldots, N(k)$) and $q_j^{(\nu)}$ is bounded then Φ has period $2\pi Q_j/\lambda_j$ in τ_j ($Q_j = \max q_j(k)$).

Suppose now ϕ is u.a.p., and put $\chi(t) = \phi(t,x_0)$. The set $M = \overline{\phi(R,\ x_0)}$ is minimal, and it is easy to prove that $m \in M \Rightarrow \phi(t,m)$ is u.a.p., with the same exponents as $\phi(t,x_0)$. Also if $\Phi(\tau_1', \tau_2',\ldots) = \Phi(\tau_1'', \tau_2'',\ldots)$ then $\tau_j'-\tau_j'' = 0 \bmod 2\pi/\lambda_j$, $j=1,2,\ldots,$ so $\Phi(\underline{\tau}') \neq \Phi(\underline{\tau}'')$ provided $0 < \left\{\begin{matrix}\tau_j' \\ \tau_j''\end{matrix}\right\} < \frac{2\pi}{\lambda_j}$, $j = 1,2,\ldots$. Hence this set in $\underline{\tau}$-space is mapped (1:1) continuously onto a subset of M, so $J = \dim$ ($\underline{\tau}$-space) $\leq n - 1$ (since a bounded minimal set in R^n is of dimension $\leq n - 1$). Thus $J(\nu) \leq n - 1$ for all ν, and it can be shown that $J = \dim M$.

If $J = n - 1$ it follows by compactification from a theorem of Kodaira and Abe on topological groups that $q_j^{(\nu)} < Q_j < \infty$ for $j = 1,2,\ldots,J$, $\nu = 1,2,\ldots,$ so ϕ is quasi-periodic. The van Dantzig solenoid gives an example of ϕ with one basic frequency in 3 dimensions which is not periodic.

For a non-autonomous u.a.p. system $\dot{x} = \psi(x,t) \sim \sum_{\nu=1}^{\infty} A_\nu(x)e^{i\Lambda_\nu^* t}$ similar methods show that there are at most $n - 1$ basic frequencies of a u.a.p. solution _additional_ to those of $\psi(x,t)$.

The results for $\dot{x} = \psi(x)$ hold also for almost periodic _flows_, and also (as G. R. Sell pointed out) for flows on a manifold V, provided that if V is a torus we admit $M = V$ so $J \leq n$ instead of $J \leq n - 1$.

REFERENCES:

[1] Cartwright, M.L. Almost periodic flows and solutions of differential equations, _Proc. L.M.S._ 17 (1967) 355-380. Corrigenda 17 (1967) 768.

[2] Cartwright, M.L. Almost periodic differential equations and almost periodic flows,

J. Diff. Equations 5 (1969) 167-181.

(11) Anosov diffeomorphisms J. Franks

We consider the problem of classifying all Anosov diffeomorphisms of compact manifolds up to topological conjugacy (see [1] or [3] for definitions). The simplest examples of Anosov diffeomorphisms are automorphisms of the n-dimensional toral group T^n whose derivatives at the identity are hyperbolic (i.e. have no eigenvalues of absolute value one). There are however, numerous examples where the manifold is not a torus (see [3]).

Piecing together results of Anosov, Smale, Sinai, and Bowen we obtain the following

THEOREM. If $f: M \to M$ is an Anosov diffeomorphism then the following properties are equivalent:

a) The non-wandering set of f, $\Omega(f)$, is all of M;

b) The set of periodic points, Per(f), is dense in M;

c) For all $x \in M$ the stable manifold $W^s(x)$ is dense in M;

d) For all $x \in$ Per(f), $W^s(x)$ is dense in M;

e) There exists an f invariant measure on M which is positive on open sets;

f) There exists a measure as in e) which is ergodic.

Definition. An Anosov diffeomorphism satisfying a) through f) will be called ergodic.

A recent result of S. Newhouse states that if either the expanding or contracting bundle of an Anosov diffeomorphism f is one-dimensional then f is ergodic. It seems reasonable to conjecture that, in fact, all Anosov diffeomorphisms are ergodic. Using Newhouse's result to weaken the hypothesis of a result in [1] we obtain

THEOREM. If $f: M^n \to M^n$ is an Anosov diffeomorphism and either the expanding or contracting bundle of f is one-dimensional then

a) M^n is homeomorphic to T^n,

b) f is topologically conjugate to a hyperbolic toral automorphism.

Without the dimension restriction we need a much stronger hypothesis.

THEOREM [2]. If $f:T^n \to T^n$ is an ergodic Anosov diffeomorphism and $f_*:H_1(T^n,R) \to H_1(T^n,R)$ is hyperbolic then f is topologically conjugate to a hyperbolic toral automorphism.

REFERENCES:

[1] Franks, J. Anosov diffeomorphisms, (to appear in *Proceedings of AMS Institute on Global Analysis, Berkeley, 1968.*)

[2] Franks, J. Anosov diffeomorphisms on tori, *Trans. A.M.S.* 145 (1969) 117-125.

[3] Smale, S. Differentiable dynamical systems, *Bull. A.M.S.* 73 (1967) 747-817.

(12) Sufficiency of jets

T. C. Kuo

Definition. An r-jet $z \in J^r(n,p)$ is *v-sufficient* if for any two realizations f, g of z, the germs of the varieties $f^{-1}(0)$ and $g^{-1}(0)$ are homeomorphic; call z *C^0-sufficient* if there exist local homeomorphisms h of R^n, h' of R^p, such that $h'f = gh$ near 0 ([1] , [2], [3]).

For a C^{r+1} mapping f, let $j^{(r)}(f) \in J^{(r)}(n,p)$ be the Taylor expansion of f at 0 up to degree $\leq r$.

Problem. Find criteria for sufficiency.

(I) In $J^r(n,1)$, we have

THEOREM (see [1], [2], and in particular [3]). For a C^∞ function $f : R^n \to R^1$, if there exist $\varepsilon > 0$, $\delta > 0$ and an integer r such that $|\text{Grad } f(x)| \geq \varepsilon |x|^{r-\delta}$ for all x near 0, then $j^{(r)}(f)$ is C^0-sufficient.

(II) In $J^r(2,1)$, the problem is completely solved in [3]. The criterion given in [3] can be simplified by a new result of Lu in [4], where he also settles a problem of Thom on parabolic umbilics.

(III) For general jets $z \in J^r(n,p)$, there are two criteria for v-sufficiency.

For $f = (f_1, \ldots, f_p) : R^n \to R^p$, and for $d > 0$, $w > 0$, the horn-neighbourhood, $H_d(f,w)$, of the variety $f^{-1}(0)$ with width w, degree d, is the set of all $x \in R^n$ satisfying $|f(x)| \leq w|x|^d$. (See [3,§3].)

THEOREM.* For a C^∞-mapping $f : R^n \to R^p$, suppose there exist integers $r_1 > 0$, $r_2 \geq 0$, and positive numbers ε, δ, w such that for all $x \in H_{r_1+r_2}(f,w)$ in a neighbourhood of 0 we have

(A) $|\text{Grad } f_i| \geq \varepsilon |x|^{r_1-\delta}$ for $1 \leq i \leq p$,

(B) $\det(a_{ij}) \geq \varepsilon |x|^{r_2}$,

where a_{ij} represents the inner product

$$|\text{Grad } f_i|^{-1} \text{Grad } f_i \cdot |\text{Grad } f_j|^{-1} \text{Grad } f_j .$$

Then $j^{(r_1+r_2)}(f)$ is v-sufficient.

The proof is similar to that of Theorem 1, [2].

Now let $z \in J^r(n,p)$ be represented by

$$f_k(x) = H_1^{(k)}(x) + \ldots + H_r^{(k)}(x) \qquad (J)$$

---

* Added in proof: This Theorem has been improved.

where $1 \leq k \leq p$. Let $V_q^{(k)}$ denote the projective variety $H_q^{(k)} = 0$, and let $SV_q^{(k)}$ denote its singular subvariety (defined by $\mathrm{Grad}\ H_q^{(k)} = 0$). If $H_q^{(k)} \equiv 0$, we set $SV_q^{(k)} = V_q^{(k)}$ ($= RP^{n-1}$). Call $u \in RP^{n-1}$ a <u>sensitive</u> point of $f_k = 0$ if $u \in SV_q^{(k)}$ for $1 \leq q \leq r$. Given k, and given u, which is not a sensitive point of $f_k = 0$, let $q = q(k)$ be the first integer such that $u \notin SV_q^{(k)}$, and let $P_k(u) = \{V_1^{(k)}, \ldots, V_q^{(k)}\}$ if $u \in V_q^{(k)}$; and $P_k(u) = \{V_1^{(k)}, \ldots, V_{q-1}^{(k)}, S\}$ otherwise, where S is the level surface $H_q^{(k)} = C$ passing through u. Let P(u) denote the disjoint union of $P_k(u)$ for $1 \leq k \leq p$. From the above theorem, we can derive

<u>THEOREM</u>. A jet $z \in J^r(n,p)$, represented by (J), is v-sufficient if the following two conditions are satisfied:

(1) For every k, and every sensitive point u of $f_k = 0$, there is an f_j ($j \leq p$) whose initial form does not vanish at u.

(2) For any u which is an insensitive point of every $f_k = 0$, each member of P(u) admits a stratification satisfying the condition (α) below; moreover, the family of strata containing u intersects transversally at u (i.e. the sum of the normal spaces at u is a direct sum; see [2]).

For a stratification $V = M_1 \cup \ldots \cup M_s$ of a hypersurface V which is defined by an equation $H(x) = 0$, <u>condition (α)</u> reads:

(α) If we are given a point $p \in M_i$, a sequence $x_m \to p$ in R^n and a sequence a_m of real numbers such that $v = \lim a_m \mathrm{Grad}\ H(x_m)$ exists, then v belongs to the normal space of M_i at p.

(IV) For C^∞-sufficiency of jets, see Mather [5] and Tougeron [6].

<u>REFERENCES</u>:

[1] Kuiper, N.H. C^1-equivalence of functions near isolated critical points, <u>Symposium</u>

on Infinite Dimensional Topology, Princeton University Press, 1968.

[2] Kuo, T.C. On C^0-sufficiency of jets of potential functions, Topology 8 (1969) 167-171.

[3] Kuo, T.C. A complete determination of C^0-sufficiency in $J^r(2,1)$, Inv. Math. 8 (1969) 226-235.

[4] Lu, Y.C. Sufficiency of jets in $J^r(2,1)$ via decomposition, (to appear).

[5] Mather, J.N. Stability of C^∞-mappings III: Finitely determined map-germs, Publ. I.H.E.S. No. 35.

[6] Tougeron, J-C. Idéaux de fonctions différentiables I, Ann. Inst. Fourier, Grenoble 18 (1968) 177-240.

(13) Asymmetric manifolds — R. Palais

(Report withdrawn)

(14) Universal unfoldings — J. Mather

(Report not received)

(15) Functional-differential systems and pattern learning

S.Grossberg

1. Introduction

The theory of embedding fields is a nonstationary prediction theory, or learning theory, described by cross-correlated flows on signed directed networks which obey suitable systems of nonlinear functional-differential equations. Mathematically speaking, one's task is to analyse globally the limiting and oscillatory behaviour of these systems.

The systems are derived from simple psychological facts such as: predicting the letter B, given the letter A, can be accomplished after practising the list AB sufficiently often. Given the systems, one perturbs them with inputs representing complicated psychological events, and compares system behaviour with analogous experimental data. A plausible neurophysiological, anatomical, and biochemical interpretation of system variables is available. Given this interpretation, the psychologically derived laws suggest physiological predictions. The psychologically derived systems also suggest a large class of functional-differential systems that are amenable to global analysis.

2. Review of Some Results

Given psychological, neurophysiological, anatomical and biochemical interpretations of the systems, their qualitative mathematical behaviour is seen to be analogous to data of the following kind. (1) Serial learning (bowing, backward learning, anchoring, chunking, all-or-none vs. gradualist learning); (2) stimulus sampling; (3) lateral inhibition (Hartline-Ratliff); (4) pattern discrimination (nonrecurrent vs. recurrent inhibition, Hubel-Wiesel cells, competitive behavioural modes, on-off fields);

(5) space-time pattern learning (avalanches of axon collaterals, command cells, ritualistic performances vs. sensitivity to feedback, suppression of background noise in the linearly ordered case and cerebellar circuitry); (6) energy-entropy dependence (minimal input entropy yields maximal energy transfer); (7) energy-learning dependence (increasing learning or input energy can decrease reaction time); (8) transmitter production and cellular control (joint pre- and post-synaptic control of transmitter production, chemical dipole in cell using ionic control by Na^{+}, K^{+}, Ca^{++}, and Mg^{++}, Mg^{++} activation of RNA, mitochondria near synaptic vesicles, stronger Ca^{++} binding than K^{+} binding, transmitter mobilization and feedback inhibition); (9) phase transitions in memory (changing network anatomy or rate parameters - such as spike velocity - can change the class of remembered events); (10) wave-particle dualism (both statistical and deterministic interpretations of the dynamics can be given, due to existence of "hidden" inhibitory interactions); (11) spatiotemporal masking and consolidation (masking of prior events by later events, such as spatially contigous light flashes, or memories in different environments); (12) Pavlovian conditioning and "practice makes perfect"; (13) memory (perfect without overt or covert practice, exponential decay, spontaneous recovery and post-tetanic potentiation, reminiscence); (14) error correction; (15) operant conditioning (internal drives, nonspecific arousal, paying attention, novelty vs. habituation, goals, facilitatory vs. incompatible behaviours).

3. Some Networks

The psychological derivation yields networks of the following type.

$$\dot{x}_i(t) = -\alpha_i x_i(t) + \Sigma_{k=1}^n [x_k(t-\tau_{ki})-\Gamma_{ki}]^+ \beta_{ki} z_{ki}^{(+)}(t)$$

$$-\Sigma_{k=1}^n [x_k(t-\sigma_{ki})-\Omega_{ki}]^+ \gamma_{ki} z_{ki}^{(-)}(t) + I_i(t), \qquad (1)$$

$$\dot{z}_{jk}^{(+)}(t) = -u_{jk}^{(+)} z_{jk}^{(+)}(t) + v_{jk}^{(+)} [x_j(t-\tau_{jk}) - \Gamma_{jk}]^+ [x_k(t)]^+, \qquad (2)$$

and

$$\dot{z}_{jk}^{(-)}(t) = -u_{jk}^{(-)} z_{jk}^{(-)}(t) + v_{jk}^{(-)} [x_j(t-\sigma_{jk}) - \Omega_{jk}]^+ [-x_k(t)]^+, \qquad (3)$$

where $[w]^+ = \max(w, 0)$, and $i, j, k = 1, 2, \ldots, n$. Here (2) can, for example, be changed to

$$\dot{z}_{jk}^{(+)}(t) = \{-u_{jk}^{(+)} + v_{jk}^{(+)} [x_k(t)]^+\} [x_j(t-\tau_{jk}) - \Gamma_{jk}]^+.$$

A theorem concerning spatial pattern learning is stated below for more general systems. Given any finite sets of indices I and J such that $I \cap J = \emptyset$ or $I = J$, consider the system

$$\dot{x}_i(t) = A(W_t,t)x_i(t) + \sum_{k\varepsilon J} B_k(W_t,t) z_{ki}(t) + \theta_i C(t)$$

and

$$\dot{z}_{ji}(t) = D_j(W_t,t) z_{ji}(t) + E_j(W_t,t) x_i(t)$$

where $i \varepsilon I$, $j \varepsilon J$, and $W = (x_i, z_{ji} : i\varepsilon I, j\varepsilon J)$. $A(W_t,t)$ is, for example, possibly a nonlinear functional of W evaluated at all times $v \leq t$, and of independent functions of t which do not depend on the index i. Moreover

1) C and all B_j and E_j are nonnegative ;
2) $\int_0^\infty B_j(W_t,t)dt = \infty$ only if $\int_0^\infty E_j(W_t,t)dt = \infty$;
3) All functionals and inputs are continuous in t;
4) The system is bounded;
5) All θ_i are nonnegative, and $\sum_{i\varepsilon I} \theta_i = 1$;
6) There exist positive constants K_1 and K_2 such that for

all $T \geq 0$, $\int_T^{T+t} C(v) \exp\left[\int_v^{T+t} A(W_\xi, \xi)d\xi\right] dv \geq K_1$ if $t \geq K_2$;

and

7) $\int_0^\infty C(v)\, dv = \infty.$

Then for arbitrary nonnegative and continuous initial data, the ratios $X_i(t) = x_i(t) [\Sigma_{k\varepsilon I} x_k(t)]^{-1}$ and $Z_{ji}(t) = z_{ji}(t) [\Sigma_{k\varepsilon I} z_{jk}(t)]^{-1}$ have limits Q_i and P_{ji}, respectively, as $t \to \infty$. Moreover, $Q_i = \theta_i$, and if $\int_0^\infty E_j(W_t, t)\, dt = \infty$, then also $P_{ji} = \theta_i$.

Space-time patterns can be learned by an "avalanche" of such systems.

REFERENCES:

[1] Grossberg, S. Embedding fields: a theory of learning with physiological implications, J. Math. Psych. 6 (1969) 209-239.

[2] " On learning, information, lateral inhibition, and transmitters, Math. Biosci. 4 (1969) 255-310.

[3] " On the serial learning of lists, Math. Biosci. 4 (1969) 201-253.

[4] " On the production and release of chemical transmitters and related topics in cellular control, J. Theoret. Biol. 22 (1969) 325-364.

[5] " Some networks that can learn, remember, and reproduce any number of complicated space-time patterns, I, J. Math. and Mech. 19 (1969) 53-91.

[6] " On learning and energy-entropy dependency in recurrent and non-recurrent signed networks, J.Stat. Phys. 1 (1969) 319-350.

[7] " On neural pattern discrimination, (to appear in J. Theoret. Biol.).

(16) Stability theory for partial differential equations P.Parks

Recently there has been a growth of interest in applying the second method of Liapunov to dynamical systems described by partial differential equations - 'distributed parameter systems' in the control engineers' language. The idea was first proposed by V.I. Zubov [7] and A.A. Movchan [1] in 1959, but practical applications have come only quite recently. As with systems of ordinary differential equations, an outstanding problem is how to find a suitable Liapunov functional.

The second method of Liapunov, as applied to the discussion of the asymptotic stability of an equilibrium point in the state or phase space of a system described by ordinary differential equations, is essentially a method of showing that the distance from the equilibrium point of a point moving along a trajectory of the disturbed system tends to zero as time increases. Although the point is described as an n-vector the stability argument is entirely in terms of scalar quantities - the Liapunov function V, its rate of change $\dot{V}$ and the distance ρ.

When a distributed system is considered it is still possible to consider scalar measures of the system disturbed from its equilibrium state. For example, a stretched string displaced z(x,t) from its equilibrium position may have as a suitable measure

$$\rho = \{\int_{x=0}^{\ell} z^2(x,t) + (\frac{\partial z}{\partial t}(x,t))^2 dx\}^{\frac{1}{2}}.$$

These measures must obey simple rules for metric space norms such as the triangle inequality.

A Liapunov function V or functional such that

$$\alpha\rho^2 \leq V \leq \beta\rho^2 , \quad \dot{V} \leq -\gamma\rho^2 , \quad \alpha, \beta, \gamma > 0$$

is sufficient to prove uniform asymptotic stability in the large, of the equilibrium point or state.

The following simple application illustrates the Liapunov functional technique.

Divergence of an aircraft wing [3].

The equation is

$$I\frac{\partial^2\theta}{\partial t^2} - \frac{\partial}{\partial y}\left(GJ\frac{\partial\theta}{\partial y}\right) = k_\theta\,\theta + k_{\dot\theta}\frac{\partial\theta}{\partial t}$$

where $\theta = 0$ at $y = 0$ and $\frac{\partial\theta}{\partial y} = 0$ at $y = \ell$.

Define $\rho^2 = \int_0^\ell \theta^2 + (\frac{\partial\theta}{\partial t})^2\, dy$ and consider

$$V = \tfrac{1}{2}\int_0^\ell GJ(\frac{\partial\theta}{\partial y})^2 + I(\frac{\partial\theta}{\partial t})^2 - k_\theta\,\theta^2\, dy$$

for which $\dot V = \int_0^\ell k_{\dot\theta}\,(\frac{\partial\theta}{\partial t})^2\, dy$, where $k_{\dot\theta} < 0$. Then $V \geq \alpha\rho^2$ if $GJ > \frac{1}{2}\ell^2 k_\theta$ on applying the Schwarz inequality

$$\{\int_a^b f(y)\; g(y)\; dy\}^2 \leq \int_a^b f^2(y)dy. \int_a^b g^2(y)dy \qquad \text{with}$$

$a = 0$, $b = y$, $f(y) = \frac{\partial\theta}{\partial y}$, $g(y) = \ell$.

This proves stability for $GJ > \frac{1}{2}\ell^2 k_\theta$, but further argument is required to justify asymptotic stability. There is an exact result that $GJ > \frac{4}{\pi^2}\,\ell^2 k_\theta$.

Hitherto construction of Liapunov functionals has been based on writing down expressions for the total kinetic and potential energy (suitably modified) or by considering integrals of quadratic forms. A more systematic scheme for construction is desirable.

One such system is to consider the differential equation as an operator L such that $L(u) = 0$. L consists of terms involving $\frac{\partial}{\partial t}$ and also space derivatives. We now differentiate $L(u)$ formally with respect to the symbol $\frac{\partial}{\partial t}$ to obtain a new operator $N(u)$. The Liapunov functional is formed from the integral with respect to space and time of the product N(u)L(u).

The terms in this product may be grouped into two groups, first a group of terms which may be integrated explicitly with respect to time and then the remaining terms which cannot be integrated in this way. The Liapunov functional V is taken to be the first group of terms. Since the product N(u)L(u) = 0 the time derivative $-\dot{V}$ is the second group of terms without integration with respect to time.

To illustrate this process let us consider the above example. The operator L(θ) is

$$I\frac{\partial^2\theta}{\partial t^2} - GJ\frac{\partial^2\theta}{\partial y^2} - k_\theta\theta - k_{\dot\theta}\frac{\partial\theta}{\partial t}.$$

The operator N(θ), differentiating partially with respect to the symbol $\frac{\partial}{\partial t}$, is

$$2I\frac{\partial\theta}{\partial t} - k_{\dot\theta}\,\theta.$$

$$\int_t\int_{y=0}^{y=\ell} L(\theta)N(\theta)\;dydt = \int_{y=0}^{\ell} I^2\left(\frac{\partial\theta}{\partial t}\right)^2 - Ik_{\dot\theta}\theta\frac{\partial\theta}{\partial t} + \tfrac{1}{2}k_{\dot\theta}^2\theta^2$$
$$- Ik_\theta\theta^2 + IGJ\left(\frac{\partial\theta}{\partial y}\right)^2\;dy + \int_t\int_{y=0}^{\ell} -Ik_{\dot\theta}\left(\frac{\partial\theta}{\partial t}\right)^2$$
$$- k_{\dot\theta}\,GJ\left(\frac{\partial\theta}{\partial y}\right)^2 + k_\theta k_{\dot\theta}\theta^2\;dydt.$$

Taking $V = \int_{y=0}^{\ell} I^2\left(\frac{\partial\theta}{\partial t}\right)^2 - Ik_{\dot\theta}\theta\frac{\partial\theta}{\partial t} + \tfrac{1}{2}\,k_{\dot\theta}^{\,2}\theta^2 - Ik_\theta\theta^2 + IGJ\left(\frac{\partial\theta}{\partial y}\right)^2\;dy$

we have $\dot{V} = -\int_{y=0}^{\ell} -Ik_{\dot\theta}\;\left(\frac{\partial\theta}{\partial t}\right)^2 - k_{\dot\theta}GJ\left(\frac{\partial\theta}{\partial y}\right)^2 + k_\theta k_{\dot\theta}\theta^2\;dy.$

This is a different functional from that quoted above, but in fact the same stability condition $GJ > \tfrac{1}{2}\ell^2 k_\theta$ is obtained on comparing the terms

$$\int_{y=0}^{\ell} GJ\left(\frac{\partial\theta}{\partial y}\right)^2\;dy \text{ and } k_\theta\int_{y=0}^{\ell}\theta^2\;dy$$

which appear both in $\dot{V}$ and V.

This procedure, originally used by Peyser [5], appears to be the most general procedure at present available for the construction of functionals. Further work on functional construction would be particularly useful. While applications to aeroelasticity [2,3] control [4] and hydrodynamics [6] have been published, further applications of this Liapunov technique seem likely to be seen in the next few years.

REFERENCES:

[1] Movchan, A.A. On Liapunov's direct method in problems of stability of elastic systems, Prikl. Mat. Meh. 23 (1959) 483-493.

[2] Parks, P.C. A stability criterion for panel flutter via the second method of Liapunov, A.I.A.A.J. 4 (1966) 175-177.

[3] Parks, P.C. Some applications of stability analysis in engineering and applied mathematics, Proc. IV. International Conference on Nonlinear Oscillations, Prague 1967 (Academia, Prague, 1968).

[4] Parks, P.C. and A.J. Pritchard. On the construction and use of Liapunov functionals, Proc. 4th Congress of IFAC, Warsaw 1969. Paper 20.5.

[5] Peyser, G. Energy integrals for the mixed problem in hyperbolic partial differential equations of higher order, J. Math. and Mech. 6 (1957) 641-653.

[6] Pritchard, A.J. A study of two of the classical problems of hydrodynamic stability by the Liapunov method, J. Inst. Math. Applic. 4 (1968) 78-93.

[7] Zubov, V.I. The methods of Liapunov and their application, Leningrad, 1959.

(17) A functional approach to stability of differential equations

J.F. Barrett

The normal theory of stability deals with transient stability, i.e. it studies the motion resulting from an assumed initial disturbance of the system. This is true both for the linear Routh-Hurwitz type of theory and for the nonlinear Lyapunov theory.

In practice it is important to study the stability of systems subjected to disturbances which act constantly for a period of time. I.G. Malkin [4] extended Lyapunov's theory to this case showing that if a dynamical system in the neighbourhood of a point of stable equilibrium (in Lyapunov's sense) is constantly perturbed by disturbances whose magnitude does not exceed a certain $\delta > 0$, then the system will not depart from the equilibrium point by more than an assigned $\varepsilon > 0$.

Malkin's result thus enables the conclusion to be made that Lyapunov stability implies stability under constantly acting disturbances. Its limitation is that it only proves stability if the disturbances are "sufficiently small" and gives no indication of how small they should be. It may happen that, while the system is stable for small disturbances, there is a critical size which, if exceeded, may result in the system responding in an unbounded manner. In practice it is important to be able to estimate the critical size when this happens.

For a number of years, the author has been interested in an approach to stability using functional analysis based directly on the concept of stability under constantly acting disturbances. This approach enables, in principle at least, such a critical size to be estimated.

A system which is subject to disturbances or forcing may be

represented by the following block diagram:

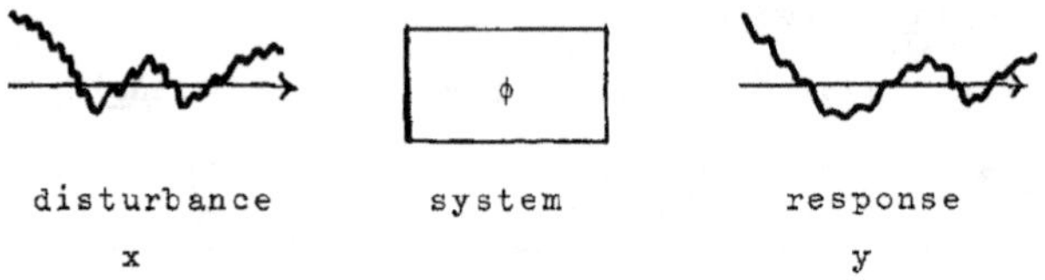

The system is thus equivalent to an operator or mapping ϕ transforming a disturbance x into a response y. Here x and y are usually vector functions of time t: $x(t)$, $y(t)$, $t \in I$.

$x = 0$ (i.e. no disturbance) usually corresponds to an equilibrium position of the system which may be taken as $y = 0$. If the system is stable under constantly acting disturbances, then a δ-neighbourhood of $x = 0$ corresponds to an ε-neighbourhood of $y = 0$. So stability under constantly acting disturbances corresponds to local continuity of ϕ at $x = 0$.

The appropriate norm is the uniform norm. Taking x and y for simplicity real variables we have

$$||x|| = \max_{t \in I} |\, x(t) \,| \qquad ||y|| = \max_{t \in I} |\, y(t) \,| \, .$$

The definition of stability under small disturbances is then that given $\varepsilon > 0$ there is a $\delta > 0$ such that if $||x|| \leq \delta$ then $||y|| \leq \varepsilon$. Analogously the system will be said to be stable under (large) disturbances not exceeding X in norm if there is a Y such that $||x|| \leq X$ implies $||y|| \leq Y$.

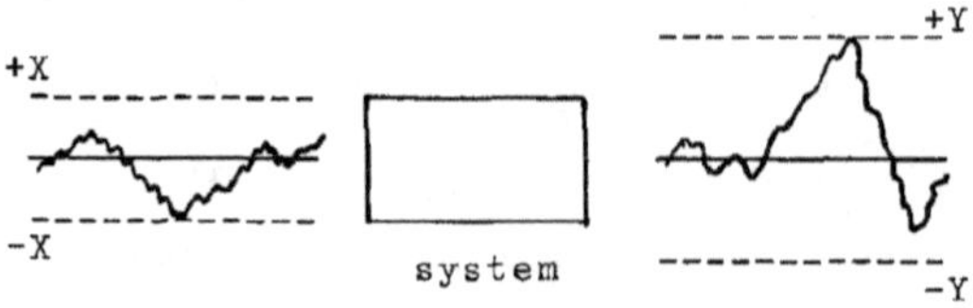

The application of these ideas to concrete problems is made possible by using Volterra expansions of the type

$$y(t) = \int_I a_1(t;u)\, x(u)\, du + \int_I \int_I a_2(t;u_1,u_2)\, x(u_1)\, x(u_2)\, du_1\, du_2 + \ldots \qquad (1)$$

which are nothing other than power series expansions in the function space.

If the inequalities

$$\int\int \ldots \int | a_n(t;u_1,u_2, \ldots, u_n)| du_1\, du_2 \ldots du_n \leq A_n, \quad t\varepsilon I,\ n=1,2,\ldots$$

are satisfied, then for $||x|| \leq X$, each of the terms of (1) does not exceed the corresponding term of the expansion

$$Y = A_1 X + A_2 X^2 + \ldots \quad = \Phi(X) \text{ say} \qquad (2)$$

which may be called a comparison series. If $||x|| < X$ then $||x|| < Y = \Phi(X)$. Consequently we may conclude that the original system is stable for disturbances for which $||x||$ does not exceed the radius of convergence of the series (2). The radius of convergence gives an estimate of the norm of the critical disturbance.

This theory would be interesting but not useful if it were not possible to estimate the radius of convergence in problems of practical importance. The author has discovered a technique which while not completely solving this problem, makes some progress to its solution for certain classes of forced nonlinear differential equations (see [1,2,3] and also [5,6]).

REFERENCES:

[1] Barrett, J.F. Thesis, Cambridge University 1958.

[2] Barrett, J.F. The use of Volterra series to find region of stability of a nonlinear differential equation, Int. J. Control 1 (1965) 209-216.

[3] Barrett, J.F. The stability of forced nonlinear systems discussed by Volterra series, Int. Conf. Appl. Mech., Quebec 1967.

[4] Makin, I.G. Theory of stability of motion, Gostekhizdat, 1952.

[5] Parente, R.B. An application of Volterra functional analysis to shunt-wound commutator machines, R.L.E., M.I.T., QPR No.76 pp. 198-208; No. 77 pp. 274-276.

[6] Parente, R.B. Functional analysis of systems characterised by nonlinear differential equations, R.L.E., M.I.T., Tech. Rep. No. 444 (1966). (D.Sc. Thesis).

(18) Numerical analysis of nonlinear oscillations — M. Urabe

One of the most powerful methods of finding periodic solutions of nonlinear periodic differential equations is the method of Galerkin. In the present talk, on the basis of [1,2,3,4] we describe the essentials of the Galerkin method and make some remarks on the results obtained by the method and on its applicability to autonomous systems.

Consider a real periodic differential system

(1) $dx/dt = X(x,t)$,

where x and $X(x,t)$ are vectors and $X(x,t+2\pi) = X(x,t)$. In the Galerkin method, corresponding to a 2π-periodic solution $\hat{x}(t)$ of (1), one seeks a Galerkin approximation, that is, a trigonometric polynomial

(2) $x_m(t) = c_1 + \sum_{n=1}^{m} (c_{2n} \sin nt + c_{2n+1} \cos nt)$

satisfying the equation

(3) $dx_m(t)/dt = P_m X[x_m(t),t]$,

where P_m denotes the truncation of the Fourier series discarding the terms of order higher than m. We can prove that under some general conditions there is always a Galerkin approximation $x_m(t)$ which approximates $\hat{x}(t)$ as accurately as we desire, if we take sufficiently large m (see [1]). Equation (3) is clearly equivalent to a system of equations of the form

(4) $F(c) = 0$,

where $c = \text{col}\,(c_1, c_2, \ldots, c_{2m+1})$. Hence a Galerkin approximation $x_m(t)$ can be found by solving (4). In many practical problems, it is possible to solve (4) numerically by the use of a computer if we apply the Newton method to (4) (see [3,4]).

Suppose that we have obtained a Galerkin approximation $\bar{x}(t)$ numerically in the above way. In this case, we usually do not yet know the existence of an exact periodic solution. However we can prove that if $\bar{x}(t)$ satisfies a certain condition (C), then (1) has indeed a periodic solution $\hat{x}(t)$ and moreover $\bar{x}(t) - \hat{x}(t)$ is within a certain bound E (see [1]). The condition (C) can be checked practically without any difficulty. Thus by checking the condition (C), we can deduce the existence of an exact periodic solution from the computed results, and moreover, we can determine an error bound E for the Galerkin approximation obtained numerically.

In the course of checking the condition (C), we compute the fundamental matrix $\Phi(t)$ of the relative first variation equation. Hence by investigating the eigenvalues of $\Phi(2\pi)$, we can study also the stability of the periodic solution.

For autonomous systems, we take a suitable smooth closed curve K and construct a moving coordinate system Σ along K (see [2]). Then with respect to Σ we have a periodic differential system whose periodic solutions correspond to closed orbits of the original autonomous system. Thus applying the Galerkin method to the periodic system obtained, we can find periodic solutions of the original autonomous system.

REFERENCES:

[1] Urabe, M. Galerkin's procedure for nonlinear periodic systems, Arch. Rat. Mech. Anal. 20 (1965) 120-152.

[2] Urabe, M. Nonlinear Autonomous Oscillations - Analytical Theory, Academic Press, New York, 1967.

[3] Urabe, M. Numerical investigation of subharmonic solutions to Duffing's equation, (to appear in Proc. V International Conference on Nonlinear Oscillations, Kiev 1969).

[4] Urabe, M. and A. Reiter. Numerical computation of nonlinear forced oscillations by Galerkin's procedure, J. Math. Analysis and Applications 14 (1966) 107-140.

(19) Dichotomies and stability theory

W.A. Coppel

The homogeneous linear differential equation

$$y' = A(t)y \tag{1}$$

is said to possess and *exponential dichotomy* (Massera and Schäffer) if it has a fundamental matrix $Y(t)$ such that

$$\begin{aligned} |Y(t)PY^{-1}(s)| &\leq Ke^{-\alpha(t-s)} \quad \text{for } t \geq s \\ |Y(t)(I-P)Y^{-1}(s)| &\leq Ke^{-\alpha(s-t)} \quad \text{for } s \geq t, \end{aligned} \tag{2}$$

where the matrix P is a projection ($P^2 = P$) and K, α are positive constants. If trace $P = k$ this means that (1) has a k-dimensional linear manifold of solutions tending to zero exponentially at $t \to \infty$, an $(n - k)$-dimensional linear manifold of solutions tending to infinity exponentially, and these manifolds are permanently transverse to one another.

THEOREM. The concept of exponential dichotomy is a good one.

A proof is given in four stages.

(i) *Stability*. If (1) has an exponential dichotomy and if

$$|g(t,x)| \leq \gamma |x|, \tag{3}$$

where $\gamma > 0$ is sufficiently small, then the null solution of the nonlinear equation

$$x' = A(t)x + g(t,x) \tag{4}$$

has stability properties similar to those of the null solution of (1). If these properties are to hold for all g satisfying (3) the assumption on (1) cannot be weakened.

(ii) Functional Analysis. (1) has an exponential dichotomy if and only if the inhomogeneous linear equation

$$x' = A(t)x + b(t) \tag{5}$$

has at least one bounded solution for every locally integrable function $b(t)$ such that $\int_t^{t+1} |b(s)|ds$ is bounded. This is deduced from results of Massera and Schäffer by an argument due to Bridgland.

(iii) Roughness. If (1) has an exponential dichotomy there exists a $\delta > 0$ such that if $|B(t)| \leq \delta$, or $\int_t^{t+1} |B(s)|\, ds \leq \delta$, for all $t \geq 0$ then the perturbed equation

$$y' = [A(t) + B(t)]y$$

also has an exponential dichotomy with the same projection P. The same holds, when $A(t)$ and $B(t)$ are bounded, if

$$\sup_{|t_2-t_1|\leq 1} \left| \int_{t_1}^{t_2} B(t)\, dt \right| \leq \delta.$$

The last result provides a new approach to Bogolyubov's method of averaging.

(iv) Practical Criterion. Let $A(t)$ be a bounded, continuous matrix function such that, for some $\alpha > 0$, $A(t)$ has k eigenvalues with real part $< -\alpha$ and $n - k$ eigenvalues with real part $> \alpha$. If

$$\sup_{|t_2-t_1|\leq 1} |A(t_2) - A(t_1)|$$

is sufficiently small then (1) has an exponential dichotomy with projection

$$P = \begin{pmatrix} I_k & 0 \\ 0 & 0 \end{pmatrix} .$$

This has applications to singular perturbation problems.

In conclusion we note analogies between exponential dichotomies and Anosov diffeomorphisms and offer a prayer that these hitherto unrelated fields might somehow interact.

REFERENCES:

[1] Chang, K.W. and W.A. Coppel. Singular perturbations of initial value problems over a finite interval, Arch. Rat. Mech. Anal. 4 (1969) 268-280.

[2] Coppel, W.A. Dichotomies and reducibility, J.Diff. Equations 3 (1967) 500-521 and 4 (1968) 386-398.

[3] Coppel, W.A. and K.J. Palmer. Averaging and integral manifolds, Bull. Austral. Math. Soc. 2 (1970) 197-222.

(20) Commuting diffeomorphisms — N. Kopell

An action of a Lie group G on a compact manifold M is a homomorphism $\rho : G \to \mathrm{Diff}(M)$. We consider here $G = Z \oplus Z$ or $G = R \oplus R$.

Actions of compact groups are very rigid: in fact if G is compact and $\bar{\rho}$ is sufficiently close to ρ there is a diffeomorphism $h : M \to M$ with $h.\rho(g) = \bar{\rho}(g).h$ for all $g \in G$. Conversely, actions of Z are very 'flexible'. For $Z \oplus Z$ and $R \oplus R$ we have some results as follows:

THEOREM 1. There exist $f,g \in \mathrm{Diff}\,(S^1)$ such that (1) $fg = gf$, (2) f, g have a common isolated fixed-point p, (3) jets $J^\infty_p(f)$, $J^\infty_p(g)$ both $= J^\infty_p(\mathrm{id})$, (4) $\bar{f},\bar{g}$ sufficiently close (C^1) to f,g and $\bar{f}\bar{g} = \bar{g}\bar{f} \Longrightarrow$ there exists $\bar{p} \in S^1$ with $\bar{f},\bar{g}$ satisfying (2),(3) with respect to $\bar{p}$.

THEOREM 2. There exist $f, g \in \mathrm{Diff}(S^1)$ such that (1) $fg = gf$, (2) $f = \mathrm{id}$ on an open set, (3) $\bar{f}, \bar{g}$ sufficiently close (C^2) to f, g and $\bar{f}\bar{g} = \bar{g}\bar{f} \Longrightarrow \bar{f} = \mathrm{id}$ on an open set.

Corollary. (Reeb, Rosenberg, Godbillon). There is a foliation of $T^2 \times I$ with all interior leaves cylinders such that any C^2-close foliation has a 'band' of cylindrical leaves.

THEOREM 3. On any M^2 there exist vector fields V, W with $[V,W] = 0$, $V = kW$ on some open set, and for any $\bar{V}$, $\bar{W}$ sufficiently C^2-close with $[\bar{V},\bar{W}] = 0$ there exists $\bar{k}$ with $\bar{V} = \bar{k}\bar{W}$ on an open set.

The proofs of Theorems 1 and 2 make use of the following

Proposition. $f : R \to R$ linear, $f \neq \mathrm{id}$, $g(C^1)$ satisfies $gf = fg \Longrightarrow g$ linear.

Proof. $g(\lambda^n x) = \lambda^n g(x) \Longrightarrow \lambda^n g'(\lambda^n x) = \lambda^n g'(x) \Longrightarrow g'(x) = g'(0)$ (all x).

For Theorem 3, we need a generalization.

Question. Does $f : R^n \to R^n$ linear ($||f|| < 1$), $gf = fg$ ($g \in C^1$) $\Longrightarrow g$ linear?

Answer. No, e.g. $f(x_1,x_2) = (\lambda x_1, \lambda^2 x_2)$ $(0 < \lambda < \ )$, $g(x_1,x_2) = (ax_1, bx_2 + cx_1^2)$.

THEOREM 4. Let $f : R^n \to R^n$ be a linear diffeomorphism, $||f|| < 1$, and let $\bar{\lambda} = \max |\lambda_i|$, $\underline{\lambda} = \min |\lambda_i|$ ($\{\lambda_i\}$ = eigenvalues). Let m = least integer such that $(\bar{\lambda})^m < \underline{\lambda}$. Then g (C^m) commutes with $f \Longrightarrow g$ is a polynomial of degree $< m$.

THEOREM 5. Let $\mathcal{A} = \{f \in \mathrm{Diff}(S^1) \mid \{g \mid gf = fg\} = \{f^n\}_{n \in Z}\}$. Then $\mathcal{A}$ is open and dense in $\mathrm{Diff}(S^1)$ (C^2 topology).

Conjecture. The same is true for general compact M.

(21) Predictions for the future of differential equations R. Abraham

Last year at the Berkeley Summer Institute on Global Analysis there were 20 talks on differential equations, of which 10 were concerned with the "yin-yang problem". Some large (yin) sets of differential equations with generic properties are known, some small (yang) sets which can be classified are known, but in general the two domains have not yet met. They began to approach each other, but

yang yin
simple ←———)————————(———→ complicated

recently progress degenerated into a sequence of conjectures and counter-examples without further approach. For guidance on future progress, the I Ching was consulted on the question : will the current program lead to a solution to the yin-yang problem - a set of differential equations both small enough to be classified and large enough to be generic? The prophesy obtained was hexagram 13 (Breakthrough) changing to 49 (Revolution), interpreted as No.

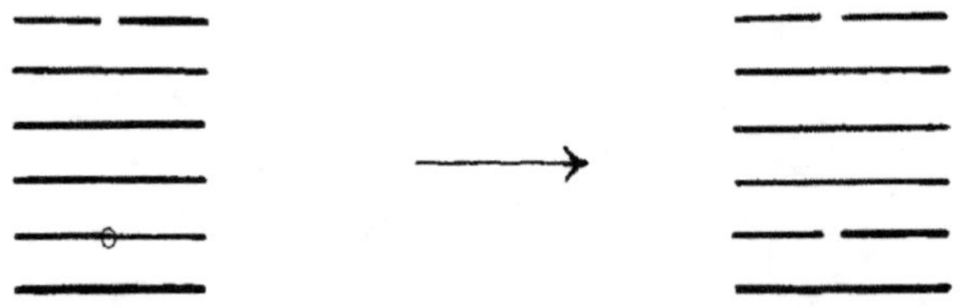

43 49

This year at the Warwick Summer School there were 22 talks of which only 3 were concerned with the yin-yang problem. The I Ching was again consulted (using for the first time, half-crowns). As it had become clear through the esoteric Buddhist principles of Karma and Transcendence that the yes-no question formerly posed was too restrictive, the question asked this time was : How will the subject evolve in the course of the next year? In view of the theories propounded by Professor Thom, it seems that in place of a steady approach of the two domains we should expect a number of bifurcations, and fruitful investigations of new domains.

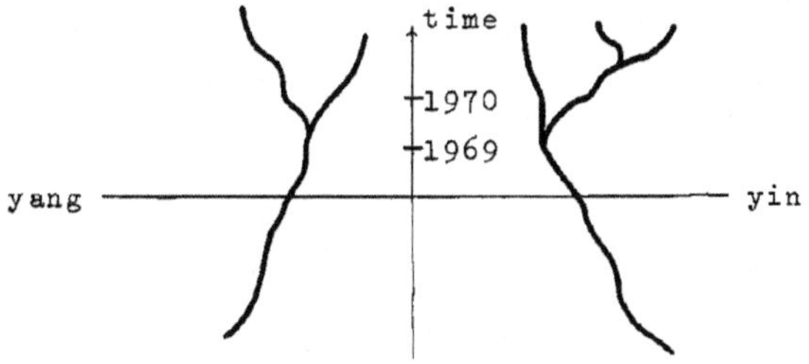

The answer to the question was given by the <u>I Ching</u> as hexagram 10 (Conduct) transforming to 11 (Peace), a clear indication of the future dominance of yin.

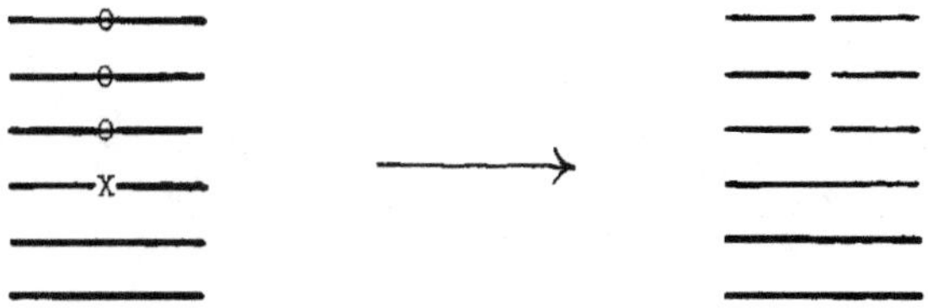

10 11

From further study of hexagram 10, the following precepts emerged:

The superior man discriminates between high and low;
One should not attempt to exceed one's own strength;
Overcome danger by going forward in time;
Resolute conduct and perseverance will lead to success;
Weigh past conduct : if it is good, then good will follow.

Thus by concentrating on what is important without being too ambitious, by continuing to work hard and with determination, and by studying closely and learning from the many counter-examples of the past, progress will be made towards peace and harmony in differential equations. Although the yin and yang aspects of differential equations will continue to oppose each other as prophesied last year, a harmonious balance between the conflicting forces may be obtained by proper study.

For Ralph

Concerned with forms we gain a healthy
disrespect for their authority, as
Dylan and Donovan mingle with
Dynamics in minds meeting answers,
the Book of Changes intruding with
equanimity in our lives, to questions
we have not yet thought to ask:
What if Ω meant more to us
than politics and death?

Michael Shub.

AUTHOR INDEX

and

LIST OF ADDRESSES

Coppel, W.A.	Department of Mathematics, Institute of Advanced Studies, Australian National University, P.O. Box 4, A.C.T. 2600, AUSTRALIA.	160
Davies, T.V.	Department of Mathematics, University of Leicester, University Road, LEICESTER.	13
Ebin, D.	Department of Mathematics, S.U.N.Y., Stony Brook, New York 11790, U.S.A.	83
Ellis, R.	School of Mathematics, University of Minnesota, Minneapolis, Minn 55455, U.S.A.	66
Epstein. D.B.A.	Mathematics Institute, University of Warwick, COVENTRY, Warwickshire CV4 7AL.	52
Franks, J.	Department of Mathematics, Northwestern University, Evanston, Ill 60201, U.S.A.	142
Furstenberg, H.	Institute of Mathematics, Hebrew University of Jerusalem, ISRAEL.	99
Gérard, R.	Université de Strasbourg, Départment de Mathématique, Rue René Descartes, 67-Strasbourg, FRANCE.	87
Godbillon, C.	Université de Strasbourg, Départemeent de Mathématique, Rue René Descartes, 67- Strasbourg, FRANCE.	9,91,104
Goodwin, B.	Department of Biological Sciences, University of Sussex, Falmer, Brighton, SUSSEX.	79
Green, L.W.	School of Mathematics, University of Minnesota, Minneapolis, Minn 55455, U.S.A.	25
Grossberg, S.	Department of Mathematics, M.I.T., Cambridge, Mass 02139, U.S.A.	147
Guckenheimer, J.	Institute for Advanced Study, Princeton, N.J. 08540, U.S.A.	45

Halanay, A.	Mathematics Institute, Cal. Grivitei 21, Bucureşti 12, ROMANIA.	106
Hirsch, M.	Department of Mathematics, University of California, Berkeley, Cal 94720, U.S.A.	90,108
Kalman, R.E.	Department of Operations Research, Stanford University, Stanford, Cal 94305, U.S.A.	138
Kopell, N.	Department of Mathematics, M.I.T., Cambridge, Mass. 02139, U.S.A.	162
Kuo, T.C.	Department of Mathematics, Manchester University, MANCHESTER.	143
Kurzweil, J.	Matematický Ústav ČSAV, Žitná 25, Praha 1, CZECHOSLOVAKIA.	9,47
Kushner, H.	Center for Dynamical Systems, Division of Applied Mathematics, Brown University, Providence, R.I. 02912, U.S.A.	129
Laudenbach, F.	École Polytechnique, Centre de Mathématique, 17 rue Descartes, Paris V, FRANCE.	116
Markus, L.	School of Mathematics, University of Minnesota, Minneapolis, Minn 55455, U.S.A.	60
Mather, J.	Department of Mathematics, Harvard University, 2, Divinity Avenue, Cambridge, Mass 02138, U.S.A.	146
Mesarović, M.D.	Systems Research Center, Case Western Reserve University, Cleveland, Ohio 44106, U.S.A.	14,15
Meyer, K.	Department of Mathematics, University of Minnesota, Minneapolis, Minn 55455.	86,128
Nohel, J.A.	Department of Mathematics, University of Wisconsin, 213 Van Vleck Hall, Madison, Wis 53706, U.S.A.	58,69

Palais, R.	Department of Mathematics, Brandeis University, Waltham, Mass 02154, U.S.A.	146
Palis, J.	I.M.P.A., Rua Luiz de Camões 68, Rio de Janeiro, BRAZIL.	40
Parks, P.	School of Engineering Science, University of Warwick, COVENTRY, Warwickshire CV4 7AL.	151
Parry, W.	Mathematics Institute, University of Warwick, COVENTRY, Warwickshire CV4 7AL.	6,36
Parthasarathy, K.	Department of Mathematics, Manchester University, MANCHESTER.	135
Phillips, A.	Department of Mathematics, S.U.N.Y., Stony Brook, New York 11790, U.S.A.	102,120
Pugh, C.	Department of Mathematics, University of California, Berkeley, Cal 94720, U.S.A.	38
Reeb, G.	Université de Strasbourg, Départment de Mathématique, Rue René Descartes, 67-Strasbourg, FRANCE.	114,124
Reinhart, B.L.	Department of Mathematics, University of Maryland, College Park, Md 20742, U.S.A.	119
Robertson, S.A.	Department of Mathematics, Southampton University, SOUTHAMPTON SO9 5NH.	127
Robinson, R.C.	Department of Mathematics, Northwestern University, Evanston, Ill 60201, U.S.A.	35
Rosenberg, H.	Départment de Mathématique, Faculté des Sciences, 91-Orsay, FRANCE.	112,133
Roussarie, R.	École Polytechnique, Laboratoire de Mathématique, 17 rue Descartes, Paris V, FRANCE.	118

Saito, T.	University of Tokyo, College of General Education (Kyoyo Gakubu), Komaba-cho, Meguro-ku, Tokyo, JAPAN.	1,18,23,56
Schmitt, B.	Université de Strasbourg, Départment de Mathématique, Rue René Descartes, 67-Strasbourg, FRANCE.	64
Schneider, C.	Department of Mathematics, Oregon State University, Corvallis, Or 97331, U.S.A.	51
Seibert, P.	Instituto Politecnico Nacional, Escuela Superior de Fisica y Matematicas, Mexico 14, D.F., MEXICO.	97
Shub, M.	Department of Mathematics, University of California, Berkeley, Calif. 94720, U.S.A.	28,39
Simon, C.	Department of Mathematics, University of California, Berkeley, Cal 94720, U.S.A.	94
Smale, S.	Department of Mathematics, University of California, Berkeley, Cal 94720, U.S.A.	33
Smith, R.A.	Department of Mathematics, University of Durham, South Road, DURHAM.	75
Urabe, M.	Research Institute for Mathematical Sciences, Kyoto University, Kyoto, 606, JAPAN.	62,158
Vrkoč, I.	Matematický Ústav ČSAV, Žitná 25, Praha 1, CZECHOSLOVAKIA.	76
Walters, P.	Mathematics Institute, University of Warwick, COVENTRY, Warwickshire CV4 7AL.	49
Weinstein, A.	Department of Mathematics, University of California, Berkeley, Cal 94720, U.S.A.	42

Weiss, B.	Institute of Mathematics, Hebrew University of Jerusalem, ISRAEL.	81
Williams, R.F.	Department of Mathematics, Northwestern University, Evanston, Ill 60201, U.S.A.	125
Wolpert, L.	Department of Biology as Applied to Medicine, Middlesex Hospital Medical School, Cleveland Street, LONDON. W.1.	85
Wood, J.	Department of Mathematics, Princeton University, Princeton, N.J. 08540, U.S.A.	110
Zeeman, E.C.	Mathematics Institute, University of Warwick, COVENTRY, Warwickshire CV4 7AL.	2

Lecture Notes in Mathematics

Bisher erschienen/Already published

Vol. 1: J. Wermer, Seminar über Funktionen-Algebren. IV, 30 Seiten. 1964. DM 3,80 / $ 1.10

Vol. 2: A. Borel, Cohomologie des espaces localement compacts d'après. J. Leray. IV, 93 pages. 1964. DM 9,– / $ 2.60

Vol. 3: J. F. Adams, Stable Homotopy Theory. Third edition. IV, 78 pages. 1969. DM 8,– / $ 2.20

Vol. 4: M. Arkowitz and C. R. Curjel, Groups of Homotopy Classes. 2nd. revised edition. IV, 36 pages. 1967. DM 4,80 / $ 1.40

Vol. 5: J.-P. Serre, Cohomologie Galoisienne. Troisième édition. VIII, 214 pages. 1965. DM 18,– / $ 5.00

Vol. 6: H. Hermes, Term Logic with Choise Operator. III, 55 pages. 1970. DM 6,– / $ 1.70

Vol. 7: Ph. Tondeur, Introduction to Lie Groups and Transformation Groups. Second edition. VIII, 176 pages. 1969. DM 14,– / $ 3.80

Vol. 8: G. Fichera, Linear Elliptic Differential Systems and Eigenvalue Problems. IV, 176 pages. 1965. DM 13,50 / $ 3.80

Vol. 9: P. L. Ivănescu, Pseudo-Boolean Programming and Applications. IV, 50 pages. 1965. DM 4,80 / $ 1.40

Vol. 10: H. Lüneburg, Die Suzukigruppen und ihre Geometrien. VI, 111 Seiten. 1965. DM 8,– / $ 2.20

Vol. 11: J.-P. Serre, Algèbre Locale. Multiplicités. Rédigé par P. Gabriel. Seconde édition. VIII, 192 pages. 1965. DM 12,– / $ 3.30

Vol. 12: A. Dold, Halbexakte Homotopiefunktoren. II, 157 Seiten. 1966. DM 12,– / $ 3.30

Vol. 13: E. Thomas, Seminar on Fiber Spaces. IV, 45 pages. 1966. DM 4,80 / $ 1.40

Vol. 14: H. Werner, Vorlesung über Approximationstheorie. IV, 184 Seiten und 12 Seiten Anhang. 1966. DM 14,– / $ 3.90

Vol. 15: F. Oort, Commutative Group Schemes. VI, 133 pages. 1966. DM 9,80 / $ 2.70

Vol. 16: J. Pfanzagl and W. Pierlo, Compact Systems of Sets. IV, 48 pages. 1966. DM 5,80 / $ 1.60

Vol. 17: C. Müller, Spherical Harmonics. IV, 46 pages. 1966. DM 5,– / $ 1.40

Vol. 18: H.-B. Brinkmann und D. Puppe, Kategorien und Funktoren. XII, 107 Seiten, 1966. DM 8,– / $ 2.20

Vol. 19: G. Stolzenberg, Volumes, Limits and Extensions of Analytic Varieties. IV, 45 pages. 1966. DM 5,40 / $ 1.50

Vol. 20: R. Hartshorne, Residues and Duality. VIII, 423 pages. 1966. DM 20,– / $ 5.50

Vol. 21: Seminar on Complex Multiplication. By A. Borel, S. Chowla, C. S. Herz, K. Iwasawa, J.-P. Serre. IV, 102 pages. 1966. DM 8,– /$ 2.20

Vol. 22: H. Bauer, Harmonische Räume und ihre Potentialtheorie. IV, 175 Seiten. 1966. DM 14,– / $ 3.90

Vol. 23: P. L. Ivănescu and S. Rudeanu, Pseudo-Boolean Methods for Bivalent Programming. 120 pages. 1966. DM 10,– / $ 2.80

Vol. 24: J. Lambek, Completions of Categories. IV, 69 pages. 1966. DM 6,80 / $ 1.90

Vol. 25: R. Narasimhan, Introduction to the Theory of Analytic Spaces. IV, 143 pages. 1966. DM 10,– / $ 2.80

Vol. 26: P.-A. Meyer, Processus de Markov. IV, 190 pages. 1967. DM 15,– / $ 4.20

Vol. 27: H. P. Künzi und S. T. Tan, Lineare Optimierung großer Systeme. VI, 121 Seiten. 1966. DM 12,– / $ 3.30

Vol. 28: P. E. Conner and E. E. Floyd, The Relation of Cobordism to K-Theories. VIII, 112 pages. 1966. DM 9,80 / $ 2.70

Vol. 29: K. Chandrasekharan, Einführung in die Analytische Zahlentheorie. VI, 199 Seiten. 1966. DM 16,80 / $ 4.70

Vol. 30: A. Frölicher and W. Bucher, Calculus in Vector Spaces without Norm. X, 146 pages. 1966. DM 12,– / $ 3.30

Vol. 31: Symposium on Probability Methods in Analysis. Chairman. D. A. Kappos.IV, 329 pages. 1967. DM 20,– / $ 5.50

Vol. 32: M. André, Méthode Simpliciale en Algèbre Homologique et Algèbre Commutative. IV, 122 pages. 1967. DM 12,– / $ 3.30

Vol. 33: G. I. Targonski, Seminar on Functional Operators and Equations. IV, 110 pages. 1967. DM 10,– / $ 2.80

Vol. 34: G. E. Bredon, Equivariant Cohomology Theories. VI, 64 pages. 1967. DM 6,80 / $ 1.90

Vol. 35: N. P. Bhatia and G. P. Szegö, Dynamical Systems. Stability Theory and Applications. VI, 416 pages. 1967. DM 24,– / $ 6.60

Vol. 36: A. Borel, Topics in the Homology Theory of Fibre Bundles. VI, 95 pages. 1967. DM 9,– / $ 2.50

Vol. 37: R. B. Jensen, Modelle der Mengenlehre. X, 176 Seiten. 1967. DM 14,– / $ 3.90

Vol. 38: R. Berger, R. Kiehl, E. Kunz und H.-J. Nastold, Differentialrechnung in der analytischen Geometrie IV, 134 Seiten. 1967 DM 12,– / $ 3.30

Vol. 39: Séminaire de Probabilités I. II, 189 pages. 1967. DM 14,– / $ 3.90

Vol. 40: J. Tits, Tabellen zu den einfachen Lie Gruppen und ihren Darstellungen. VI, 53 Seiten. 1967. DM 6.80 / $ 1.90

Vol. 41: A. Grothendieck, Local Cohomology. VI, 106 pages. 1967. DM 10,– / $ 2.80

Vol. 42: J. F. Berglund and K. H. Hofmann, Compact Semitopological Semigroups and Weakly Almost Periodic Functions. VI, 160 pages. 1967. DM 12,– / $ 3.30

Vol. 43: D. G. Quillen, Homotopical Algebra. VI, 157 pages. 1967. DM 14,– / $ 3.90

Vol. 44: K. Urbanik, Lectures on Prediction Theory. IV, 50 pages. 1967. DM 5,80 / $ 1.60

Vol. 45: A. Wilansky, Topics in Functional Analysis. VI, 102 pages. 1967. DM 9,60 / $ 2.70

Vol. 46: P. E. Conner, Seminar on Periodic Maps.IV, 116 pages. 1967. DM 10,60 / $ 3.00

Vol. 47: Reports of the Midwest Category Seminar I. IV, 181 pages. 1967. DM 14,80 / $ 4.10

Vol. 48: G. de Rham, S. Maumary et M. A. Kervaire, Torsion et Type Simple d'Homotopie. IV, 101 pages. 1967. DM 9,60 / $ 2.70

Vol. 49: C. Faith, Lectures on Injective Modules and Quotient Rings. XVI, 140 pages. 1967. DM 12,80 / $ 3.60

Vol. 50: L. Zalcman, Analytic Capacity and Rational Approximation. VI, 155 pages. 1968. DM 13.20 / $ 3.70

Vol. 51: Séminaire de Probabilités II. IV, 199 pages. 1968. DM 14,– / $ 3.90

Vol. 52: D. J. Simms, Lie Groups and Quantum Mechanics. IV, 90 pages. 1968. DM 8,– / $ 2.20

Vol. 53: J. Cerf, Sur les difféomorphismes de la sphère de dimension trois (Γ_4= O). XII, 133 pages. 1968. DM 12,– / $ 3.30

Vol. 54: G. Shimura, Automorphic Functions and Number Theory. VI, 69 pages. 1968. DM 8,– / $ 2.20

Vol. 55: D. Gromoll, W. Klingenberg und W. Meyer, Riemannsche Geometrie im Großen. VI, 287 Seiten. 1968. DM 20,– / $ 5.50

Vol. 56: K. Floret und J. Wloka, Einführung in die Theorie der lokalkonvexen Räume. VIII, 194 Seiten. 1968. DM 16,– / $ 4.40

Vol. 57: F. Hirzebruch und K. H. Mayer, O (n)-Mannigfaltigkeiten, exotische Sphären und Singularitäten. IV, 132 Seiten. 1968. DM 10,80/ $ 3.00

Vol. 58: Kuramochi Boundaries of Riemann Surfaces. IV, 102 pages. 1968. DM 9,60 / $ 2.70

Vol. 59: K. Jänich, Differenzierbare G-Mannigfaltigkeiten. VI, 89 Seiten. 1968. DM 8,– / $ 2.20

Vol. 60: Seminar on Differential Equations and Dynamical Systems. Edited by G. S. Jones. VI, 106 pages. 1968. DM 9,60 / $ 2.70

Vol. 61: Reports of the Midwest Category Seminar II. IV, 91 pages. 1968. DM 9,60 / $ 2.70

Vol. 62: Harish-Chandra, Automorphic Forms on Semisimple Lie Groups X, 138 pages. 1968. DM 14,– / $ 3.90

Vol. 63: F. Albrecht, Topics in Control Theory. IV, 65 pages. 1968. DM 6,80 / $ 1.90

Vol. 64: H. Berens, Interpolationsmethoden zur Behandlung von Approximationsprozessen auf Banachräumen. VI, 90 Seiten. 1968. DM 8,– / $ 2.20

Vol. 65: D. Kölzow, Differentiation von Maßen. XII, 102 Seiten. 1968. DM 8,– / $ 2.20

Vol. 66: D. Ferus, Totale Absolutkrümmung in Differentialgeometrie und -topologie. VI, 85 Seiten. 1968. DM 8,– / $ 2.20

Vol. 67: F. Kamber and P. Tondeur, Flat Manifolds. IV, 53 pages. 1968. DM 5,80 / $ 1.60

Vol. 68: N. Boboc et P. Mustață, Espaces harmoniques associés aux opérateurs différentiels linéaires du second ordre de type elliptique. VI, 95 pages. 1968. DM 8,60 / $ 2.40

Vol. 69: Seminar über Potentialtheorie. Herausgegeben von H. Bauer. VI, 180 Seiten. 1968. DM 14,80 / $ 4.10

Vol. 70: Proceedings of the Summer School in Logic. Edited by M. H. Löb. IV, 331 pages. 1968. DM 20,– / $ 5.50

Vol. 71: Séminaire Pierre Lelong (Analyse), Année 1967 – 1968. VI, 190 pages. 1968. DM 14,– / $ 3.90

Vol. 72: The Syntax and Semantics of Infinitary Languages. Edited by J. Barwise. IV, 268 pages. 1968. DM 18,– / $ 5.00

Vol. 73: P. E. Conner, Lectures on the Action of a Finite Group. IV, 123 pages. 1968. DM 10,– / $ 2.80

Vol. 74: A. Fröhlich, Formal Groups. IV, 140 pages. 1968. DM 12,– / $ 3.30

Vol. 75: G. Lumer, Algèbres de fonctions et espaces de Hardy. VI, 80 pages. 1968. DM 8,– / $ 2.20

Vol. 76: R. G. Swan, Algebraic K-Theory. IV, 262 pages. 1968. DM 18,– / $ 5.00

Vol. 77: P.-A. Meyer, Processus de Markov: la frontière de Martin. IV, 123 pages. 1968. DM 10,– / $ 2.80

Vol. 78: H. Herrlich, Topologische Reflexionen und Coreflexionen. XVI, 166 Seiten. 1968. DM 12,– / $ 3.30

Vol. 79: A. Grothendieck, Catégories Cofibrées Additives et Complexe Cotangent Relatif. IV, 167 pages. 1968. DM 12,– / $ 3.30

Vol. 80: Seminar on Triples and Categorical Homology Theory. Edited by B. Eckmann. IV, 398 pages. 1969. DM 20,– / $ 5.50

Vol. 81: J.-P. Eckmann et M. Guenin, Méthodes Algébriques en Mécanique Statistique. VI, 131 pages. 1969. DM 12,– / $ 3.30

Vol. 82: J. Wloka, Grundräume und verallgemeinerte Funktionen. VIII, 131 Seiten. 1969. DM 12,– / $ 3.30

Vol. 83: O. Zariski, An Introduction to the Theory of Algebraic Surfaces. IV, 100 pages. 1969. DM 8,– / $ 2.20

Vol. 84: H. Lüneburg, Transitive Erweiterungen endlicher Permutationsgruppen. IV, 119 Seiten. 1969. DM 10,– / $ 2.80

Vol. 85: P. Cartier et D. Foata, Problèmes combinatoires de commutation et réarrangements. IV, 88 pages. 1969. DM 8,– / $ 2.20

Vol. 86: Category Theory, Homology Theory and their Applications I. Edited by P. Hilton. VI, 216 pages. 1969. DM 16,– / $ 4.40

Vol. 87: M. Tierney, Categorical Constructions in Stable Homotopy Theory. IV, 65 pages. 1969. DM 6,– / $ 1.70

Vol. 88: Séminaire de Probabilités III. IV, 229 pages. 1969. DM 18,– / $ 5.00

Vol. 89: Probability and Information Theory. Edited by M. Behara, K. Krickeberg and J. Wolfowitz. IV, 256 pages. 1969. DM 18,– / $ 5.00

Vol. 90: N. P. Bhatia and O. Hajek, Local Semi-Dynamical Systems. II, 157 pages. 1969. DM 14,– / $ 3.90

Vol. 91: N. N. Janenko, Die Zwischenschrittmethode zur Lösung mehrdimensionaler Probleme der mathematischen Physik. VIII, 194 Seiten. 1969. DM 16,80 / $ 4.70

Vol. 92: Category Theory, Homology Theory and their Applications II. Edited by P. Hilton. V, 308 pages. 1969. DM 20,– / $ 5.50

Vol. 93: K. R. Parthasarathy, Multipliers on Locally Compact Groups. III, 54 pages. 1969. DM 5,60 / $ 1.60

Vol. 94: M. Machover and J. Hirschfeld, Lectures on Non-Standard Analysis. VI, 79 pages. 1969. DM 6,– / $ 1.70

Vol. 95: A. S. Troelstra, Principles of Intuitionism. II, 111 pages. 1969. DM 10,– / $ 2.80

Vol. 96: H.-B. Brinkmann und D. Puppe, Abelsche und exakte Kategorien, Korrespondenzen. V, 141 Seiten. 1969. DM 10,– / $ 2.80

Vol. 97: S. O. Chase and M. E. Sweedler, Hopf Algebras and Galois theory. II, 133 pages. 1969. DM 10,– / $ 2.80

Vol. 98: M. Heins, Hardy Classes on Riemann Surfaces. III, 106 pages. 1969. DM 10,– / $ 2.80

Vol. 99: Category Theory, Homology Theory and their Applications III. Edited by P. Hilton. IV, 489 pages. 1969. DM 24,– / $ 6.60

Vol. 100: M. Artin and B. Mazur, Etale Homotopy. II, 196 Seiten. 1969. DM 12,– / $ 3.30

Vol. 101: G. P. Szegö et G. Treccani, Semigruppi di Trasformazioni Multivoche. VI, 177 pages. 1969. DM 14,– / $ 3.90

Vol. 102: F. Stummel, Rand- und Eigenwertaufgaben in Sobolewschen Räumen. VIII, 386 Seiten. 1969. DM 20,– / $ 5.50

Vol. 103: Lectures in Modern Analysis and Applications I. Edited by C. T. Taam. VII, 162 pages. 1969. DM 12,– / $ 3.30

Vol. 104: G. H. Pimbley, Jr., Eigenfunction Branches of Nonlinear Operators and their Bifurcations. II, 128 pages. 1969. DM 10,– / $ 2.80

Vol. 105: R. Larsen, The Multiplier Problem. VII, 284 pages. 1969. DM 18,– / $ 5.00

Vol. 106: Reports of the Midwest Category Seminar III. Edited by S. Mac Lane. III, 247 pages. 1969. DM 16,– / $ 4.40

Vol. 107: A. Peyerimhoff, Lectures on Summability. III, 111 pages. 1969. DM 8,– / $ 2.20

Vol. 108: Algebraic K-Theory and its Geometric Applications. Edited by R. M. F. Moss and C. B. Thomas. IV, 86 pages. 1969. DM 6,– / $ 1.70

Vol. 109: Conference on the Numerical Solution of Differential Equations. Edited by J. Ll. Morris. VI, 275 pages. 1969. DM 18,– / $ 5.00

Vol. 110: The Many Facets of Graph Theory. Edited by G. Chartrand and S. F. Kapoor. VIII, 290 pages. 1969. DM 18,– / $ 5.00

Vol. 111: K. H. Mayer, Relationen zwischen charakteristischen Zahlen. III, 99 Seiten. 1969. DM 8,– / $ 2.20

Vol. 112: Colloquium on Methods of Optimization. Edited by N. N. Moiseev. IV, 293 pages. 1970. DM 18,– / $ 5.00

Vol. 113: R. Wille, Kongruenzklassengeometrien. III, 99 Seiten. 1970. DM 8,– / $ 2.20

Vol. 114: H. Jacquet and R. P. Langlands, Automorphic Forms on GL (2). VII, 548 pages. 1970. DM 24,– / $ 6.60

Vol. 115: K. H. Roggenkamp and V. Huber-Dyson, Lattices over Orders I. XIX, 290 pages. 1970. DM 18,– / $ 5.00

Vol. 116: Séminaire Pierre Lelong (Analyse) Année 1969. IV, 195 pages. 1970. DM 14,– / $ 3.90

Vol. 117: Y. Meyer, Nombres de Pisot, Nombres de Salem et Analyse Harmonique. 63 pages. 1970. DM 6,– / $ 1.70

Vol. 118: Proceedings of the 15th Scandinavian Congress, Oslo 1968. Edited by K. E. Aubert and W. Ljunggren. IV, 162 pages. 1970. DM 12,– / $ 3.30

Vol. 119: M. Raynaud, Faisceaux amples sur les schémas en groupes et les espaces homogènes. III, 219 pages. 1970. DM 14,– / $ 3.90

Vol. 120: D. Siefkes, Büchi's Monadic Second Order Successor Arithmetic. XII, 130 Seiten. 1970. DM 12,– / $ 3.30

Vol. 121: H. S. Bear, Lectures on Gleason Parts. III, 47 pages. 1970. DM 6,–/$ 1.70

Vol. 122: H. Zieschang, E. Vogt und H.-D. Coldewey, Flächen und ebene diskontinuierliche Gruppen. VIII, 203 Seiten. 1970. DM 16,– / $ 4.40

Vol. 123: A. V. Jategaonkar, Left Principal Ideal Rings. VI, 145 pages. 1970. DM 12,– / $ 3.30

Vol. 124: Séminare de Probabilités IV. Edited by P. A. Meyer. IV, 282 pages. 1970. DM 20,– / $ 5.50

Vol. 125: Symposium on Automatic Demonstration. V, 310 pages. 1970. DM 20,– / $ 5.50

Vol. 126: P. Schapira, Théorie des Hyperfonctions. XI, 157 pages. 1970. DM 14,– / $ 3.90

Vol. 127: I. Stewart, Lie Algebras. IV, 97 pages. 1970. DM 10,– / $ 2.80

Vol. 128: M. Takesaki, Tomita's Theory of Modular Hilbert Algebras and its Applications. II, 123 pages. 1970. DM 10,– / $ 2.80

Vol. 129: K. H. Hofmann, The Duality of Compact Semigroups and C*- Bigebras. XII, 142 pages. 1970. DM 14,– / $ 3.90

Vol. 130: F. Lorenz, Quadratische Formen über Körpern. II, 77 Seiten. 1970. DM 8,– / $ 2.20

Vol. 131: A Borel et al., Seminar on Algebraic Groups and Related Finite Groups. VII, 321 pages. 1970. DM 22,– / $ 6.10

Vol. 132: Symposium on Optimization. III, 348 pages. 1970. DM 22,– / $ 6.10

Vol. 133: F. Topsøe, Topology and Measure. XIV, 79 pages. 1970. DM 8,– / $ 2.20

Vol. 134: L. Smith, Lectures on the Eilenberg-Moore Spectral Sequence. VII, 142 pages. 1970. DM 14,– / $ 3.90

Vol. 135: W. Stoll, Value Distribution of Holomorphic Maps into Compact Complex Manifolds. II, 267 pages. 1970. DM 18,– / $ 5.00

Vol. 136: M. Karoubi et al., Séminaire Heidelberg-Saarbrücken-Strasbourg sur la K-Théorie. IV, 264 pages. 1970. DM 18,– / $ 5.00

Vol. 137: Reports of the Midwest Category Seminar IV. Edited by S. MacLane. III, 139 pages. 1970. DM 12,– / $ 3.30

Vol. 138: D. Foata et M. Schützenberger, Théorie Géométrique des Polynômes Eulériens. V, 94 pages. 1970. DM 10,– / $ 2.80

Vol. 139: A. Badrikian, Séminaire sur les Fonctions Aléatoires Linéaires et les Mesures Cylindriques. VII, 221 pages. 1970. DM 18,– / $ 5.00

Vol. 140: Lectures in Modern Analysis and Applications II. Edited by C. T. Taam. VI, 119 pages. 1970. DM 10,– / $ 2.80

Vol. 141: G. Jameson, Ordered Linear Spaces. XV, 194 pages. 1970. DM 16,– / $ 4.40

Vol. 142: K. W. Roggenkamp, Lattices over Orders II. V, 388 pages. 1970. DM 22,– / $ 6.10

Vol. 143: K. W. Gruenberg, Cohomological Topics in Group Theory. XIV, 275 pages. 1970. DM 20,– / $ 5.50

Vol. 144: Seminar on Differential Equations and Dynamical Systems, II. Edited by J. A. Yorke. VIII, 268 pages. 1970. DM 20,– / $ 5.50

Vol. 145: E. J. Dubuc, Kan Extensions in Enriched Category Theory. XVI, 173 pages. 1970. DM 16,– / $ 4.40

Vol. 146: A. B. Altman and S. Kleiman, Introduction to Grothendieck Duality Theory. II, 192 pages. 1970. DM 18,– / $ 5.00

Vol. 147: D. E. Dobbs, Cech Cohomological Dimensions for Commutative Rings. VI, 176 pages. 1970. DM 16,– / $ 4.40

Vol. 148: R. Azencott, Espaces de Poisson des Groupes Localement Compacts. IX, 141 pages. 1970. DM 14,– / $ 3.90